OXFORD BIOLOGY PRIMERS

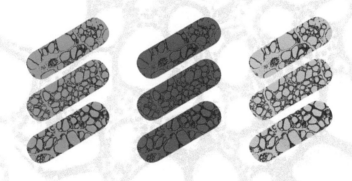

Discover more in the series at
www.oxfordtextbooks.co.uk/obp

Published in partnership with the Royal Society of Biology

ANIMAL DEVELOPMENTAL BIOLOGY

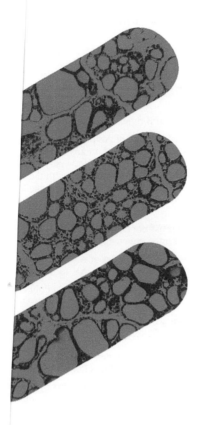

DEVELOPMENTAL BIOLOGY

s, Evolution, and Ageing

Julia Paxson

Edited by Ann Fullick

ey, Gill Hickman, Sue Howarth, and Hilary Otter

OXFORD
UNIVERSITY PRESS

Great Clarendon Street, Oxford, OX2 6DP,
United Kingdom

Oxford University Press is a department of the University of Oxford.
It furthers the University's objective of excellence in research, scholarship,
and education by publishing worldwide. Oxford is a registered trade mark of
Oxford University Press in the UK and in certain other countries

Published in the United States of America by Oxford University Press
198 Madison Avenue, New York, NY 10016, United States of America

British Library Cataloguing in Publication Data

Data available

Library of Congress Control Number: 2023946447

ISBN 978-0-19-886913-9

Printed in the UK by
Ashford Colour Press Ltd, Gosport, Hampshire

MIX
Paper from
responsible sources
FSC® C011748

PREFACE

Welcome to the Oxford Biology Primers

There has never been a more exciting time to be a biologist. Not only do we understand more about the biological world than ever before, but we're using that understanding in ever more creative and valuable ways.

Our understanding of the way our genes work is being used to explore new ways to treat disease; our understanding of ecosystems is being used to explore more effective ways to protect the diversity of life on Earth; our understanding of plant science is being used to explore more sustainable ways to feed a growing human population.

The repeated use of the word 'explore' here is no accident. The study of biology is, at heart, an exploration. We have written the Oxford Biology Primers to encourage you to explore biology for yourself—to find out more about what scientists at the cutting edge of the subject are researching, and the biological problems they're trying to solve.

Throughout the series, we use a range of features to help you see topics from different perspectives.

Scientific approach panels help you understand a little more about 'how we know what we know'—that is, the research that has been carried out to reveal our current understanding of the science described in the text, and the methods and approaches scientists have used when carrying out that research.

Case studies explore how a particular concept is relevant to our everyday life, or provide an intimate picture of one aspect of the science described.

The bigger picture panels help you think about some of the issues and challenges associated with the topic under discussion—for example, ethical considerations, or wider impacts on society.

More than anything, however, we hope this series will reveal to you, its readers, that biology is awe-inspiring, both in its variety and its intricacy, and will drive you forward to explore the subject further for yourself.

ACKNOWLEDGEMENTS

My most heartfelt thanks to all the students who have inspired me to write and to become a better teacher. Thanks also to Ann Fullick for her editorial input and pep talks during the process of writing this book. But above all else, my never-ending gratitude to my wife Paula, sons Jamis and William, and parents Pat and Dean, who have all given me such fantastic encouragement over the years. Together, you make me the person that I am.

CONTENTS

1 WHY AREN'T WE ALL WORMS?

The answer to the title of this chapter—why aren't we all worms?—is to be found in the study of developmental biology. But what is developmental biology? There are lots of answers to this question, depending on which aspect of this amazing area of biology appeals to you most. For example, cell biologists, geneticists, evolutionary biologists, and biochemists all have different ways of looking at the very interdisciplinary process that is developmental biology.

One definition of developmental biology is:

The study of the processes by which a single cell—the fertilized egg—divides to form populations of cells that communicate with each other and become highly specialized to form a complex multicellular organism, which can often repair itself and also often eventually dies.

We study developmental biology to understand the **morphological**, cellular, and molecular mechanisms that drive the growth, organization, and repair of complex multicellular organisms. The same mechanisms that occur during embryo development (see Figure 1.1) are also important in cancer biology, adult tissue repair, stem cell biology, regeneration, and ageing.

We also study developmental biology to understand evolution—and study evolution to understand developmental biology. This integrated approach is known as **evo-devo** (evolutionary developmental biology).

Where does this leave us? If you are interested in any of these areas of biology, or when you need healthcare at some stage in your life, understanding the fundamental principles of developmental biology will stand you in good stead. Not only will understanding developmental biology help you become a better-informed scientist and individual, but it will also help inform life decisions that you make.

Developmental biology is complex and fascinating in all types of organisms, but we cannot hope to do them all justice in one short primer. So here, we will focus mainly on animals—but the same processes take place in plants and organisms from all the kingdoms of the living world.

The paradox of the genetic toolkit

The genetic instructions for organisms from algae to elephants contain a paradox. They have many elements in common across thousands of species, yet their instructions are also specific to each individual organism. Let's find out how these apparent contradictions are resolved.

How well is genetic function conserved across animal species?

The small nematode worm *Caenorhabditis elegans* (*C. elegans*) is about 1mm long, has 959 somatic cells (normal body cells) and has a life span of 2–3 weeks. In comparison, the global average equivalents for a human are 159.5 cm for females and 1.71 cm for males, with global lifespans of 75.6 and 70.8 years respectively and approximately 37.2 trillion somatic cells. With these differences in mind, it is staggering to consider that about 80 per cent of the genes of *C. elegans* have equivalent (homologous) genes in humans. There are a small number of genes that encode nematode-specific proteins not found in humans, and conversely a larger number of genetic variants that encode proteins found in humans, but not in nematodes. However, given the vast differences in our respective body plans (see Figure 1.3), these genetic similarities are quite remarkable. This phenomenon is known as the evolutionary conservation of genetic function, and is a critical element both in animal evolution and in animal embryo development. Since a very similar genetic toolkit is used to create both a worm body and a human body, this begs the question—with such a conserved genetic toolkit, how is the clear diversity in body form generated?

One important clue is that less than 2 per cent of the DNA in the animal genome encodes the information needed to generate the proteins that control the reactions of cellular life. What does the remaining 98 per cent of our DNA do? For many years, we called it junk DNA—remnants from our evolutionary past perhaps. But it is now clear that these vast stretches of so-called non-coding

Figure 1.3 *Caenorhabditis elegans* and *Homo sapiens*—spot the difference!

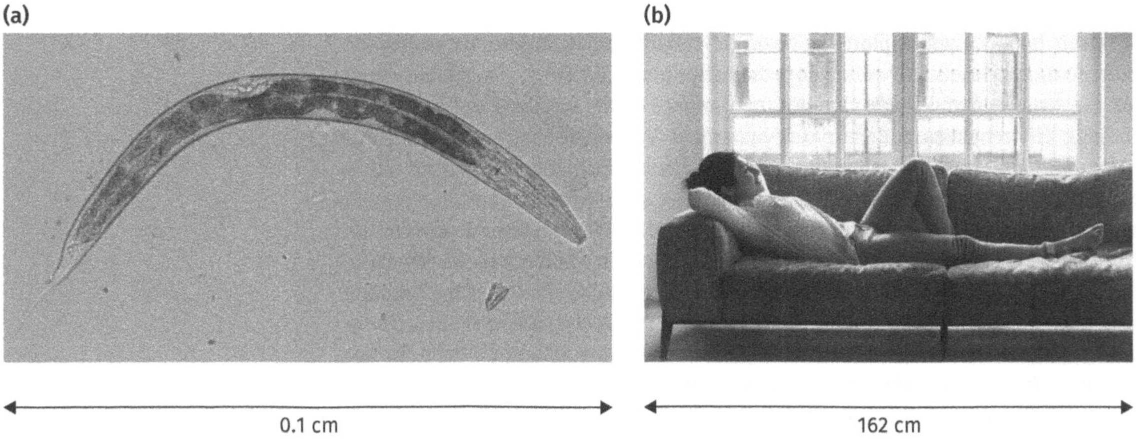

(a)   **(b)**

|◄─────────────── 0.1 cm ───────────────►| |◄─────────────── 162 cm ───────────────►|

(a): lostkabab/Shutterstock; (b): © iStock/fizkes

DNA have a vital role in our cells. This non-coding genomic **regulatory DNA** is used in a variety of ways to regulate which bits of the information encoded in the genes will be decoded to make proteins in the cell. This control is brought about through the process of **differential gene expression**—the processes that determine which genes or parts of genes are actively transcribed and turned into proteins. The particular proteins constructed using the information encoded in the DNA will determine cellular functions. Different proteins result in different cell capabilities. For example, although each cell in an individual animal carries the same DNA, liver cells must have **enzymes** such as alcohol dehydrogenase to break down or otherwise detoxify toxins, while nerve cells need neurotransmitter proteins that will allow them to transmit information across synapses. These differences are all controlled by regulating which proteins are made in each of the different cell types within the body.

Differential gene expression within a cell can occur at many different levels along the flow of genetic information from genomic DNA to protein production. The crucial two-step process, by which DNA is transcribed to give RNA, and the information in the RNA is translated at the ribosomes to produce proteins is known as the **central dogma of molecular biology**. Figure 1.4 summarizes the process of protein synthesis in eukaryotic cells. Notice that even though

Figure 1.4 Protein synthesis—from gene to protein in the cell. Specific genes are transcribed based on which parts of the genomic sequence are accessible and what combinations of transcription factors are present (1). Those mRNA sequences are then exported from the nucleus and translated by ribosomes (2). Further post-translational processing is then often necessary to produce the functional protein (3).

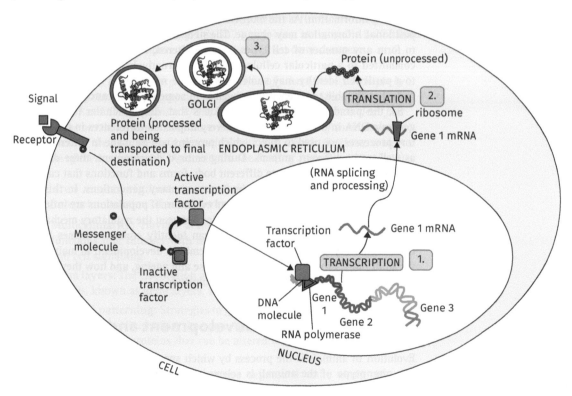

Julia Paxson

All of these themes are rooted in innovations and alterations made during embryo development. By studying a diverse range of animals that exhibit a variety of these developmentally important themes, we can explore the role of evolution in driving these themes. For example, scientists think that embryo-like fossils from the Ediacaran Doushantuo Formation (~600 million years old) shown in Figure 1.5 were early multicellular animals, complete with differential

Figure 1.5 These fossils from around 600 million years ago need experts to interpret them—but scientists think they are early, true multicellular animals. In the top panel, reconstructions of these fossils show common embryo-like patterns of cell division. The bottom panels show fossil examples of different embryo-like organisms found from this period.

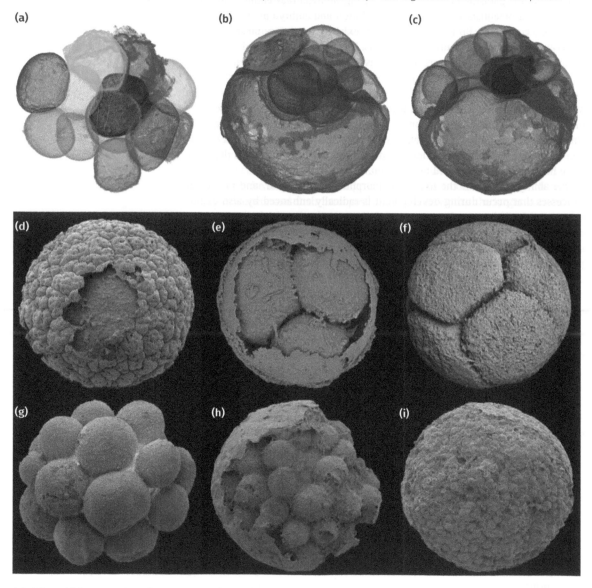

(a), (b), and (c): Dr Zongjun Yin, Nanjing Institute of Geology and Palaeontology, Chinese Academy of Sciences; (d), (e), (f), (g), (h), and (i): Cunningham, J. A., Vargas, K., Yin, Z., Bengtson, S., and Donoghue, P. C. (2017). 'The Weng'an Biota (Doushantuo Formation): An Ediacaran Window on Soft-Bodied and Multicellular Microorganisms'. *Journal of the Geological Society* 174/5 (2017), 793–802

cell adhesion between groups of morphologically distinct differentiated cells, including specialized germ cells separated from, but surrounded by, somatic cells.

Multicellularity

As illustrated by the early embryo-like fossils from the Ediacaran Doushantuo Formation, one strategy that gave some organisms an evolutionary advantage was to become multicellular. To generate a multicellular organism, newly divided cells must stick together. This might sound intuitive, but certainly would not have happened in an ancient single-celled organism that used simple cell division to reproduce. At some point in ancient history, a variation in how cells interact with one another meant that instead of two individual daughter unicellular animals being produced, the two daughter cells stuck together, creating the first multicellular animal. These remarkable adhesion events repeated, eventually giving rise to conglomerates of cells that developed specialized functions, all working together as a collective.

Phylogenomic studies from present-day single-cell eukaryotic organisms, such as choanoflagellates, have revealed that these unicellular cousins of multicellular animals (see Figure 1.6) have a large array of the genes required for multicellular development. Even though these are single-cell organisms, these genes code for characteristics commonly seen in multicellular organisms, including cell adhesion, cell signalling and regulation of differential gene expression—leading scientists to consider that choanoflagellates and multicellular animals share a common unicellular eukaryotic ancestor that had the capacity for traits needed to eventually become multicellular.

As multicellularity evolved, newly divided cells developed the capacity to have different degrees of selective adhesion between them, so that some groups of cells were more likely to stick together than others. As the number of cells involved increased, this ability to stick together evolved further, so that it allowed not only for cell aggregation but also the movement of cells relative to each other.

Figure 1.6 Genetic information about cell adhesion and cell movements has led scientists to propose the relationships between early groups of organisms.

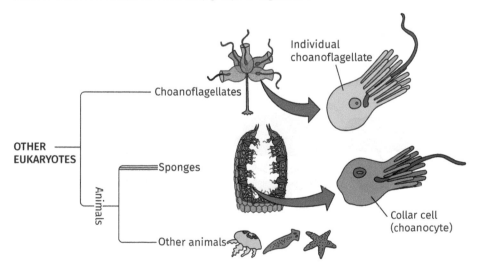

of being geographically situated in unique clusters on the same chromosome, ordered in the same way that they are expressed in the embryo. Hox genes that are expressed in the most **anterior** (towards the head) regions of the embryo are located on the 5' (upstream) end of the chromosome, while Hox genes that are expressed in the more **posterior** (tail-end) regions of the embryo are located on the 3' (downstream) end of the chromosome. While this might seem like a logical way to arrange genes that are expressed in particular patterns within the developing animal, it does not happen with any other gene series. Moreover, this remarkable arrangement, known as **collinearity**, is conserved among a very diverse range of animals (see Figure 1.9).

Figure 1.9 This illustration highlights two important features of Hox genes. First, that they are physically located along the chromosome in the same order that they are expressed (anterior to posterior) along the main body axis of the embryo. Second, that this pattern of Hox gene expression is highly conserved across diverse animal species.

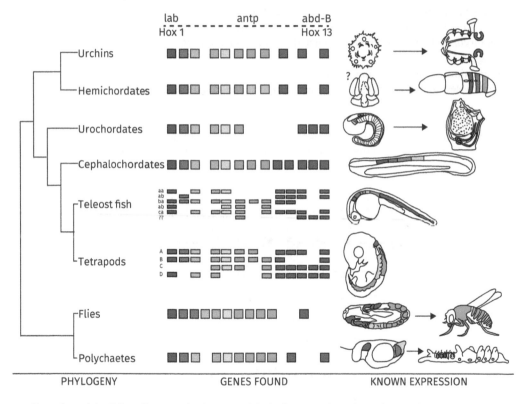

Swalla, B. (2006). 'Building divergent body plans with similar genetic pathways'. *Heredity* 97, 235–243. https://doi.org/10.1038/sj.hdy.6800872

Segmentation

There are also other aspects of animal embryo development that are well-suited to generating a wide range of phenotypes that may be targeted by evolutionary selection processes. For example, several important **clades** of animals,

Case study 1.1
Building diverse bodies using Hox genes

Hox genes were first discovered by Thomas Hunt Morgan and his students in the 1920s in fruit flies. They discovered two clusters of genes that they named the **Antennapedia** and **Ultrabithorax** complexes. Expression of a single Hox gene acts as a 'master switch' to activate an group of down-stream target genes that are necessary to specify a regional identity in the body, such as wings, halteres, or legs in fruit flies, or distinct regional vertebral structures in vertebrates. So, in the *Drosophila* embryo, the Hox gene Antennapedia usually controls much of the thorax development, and par-ticularly the growth of the second pair of legs (see Figure A).

Because they are so powerful in driving the body plan, Hox gene mutations are striking. They result in huge varia-tions in the body—known as **homeotic transformations**—which completely change the final structural and functional identity of a specific body segment. In *Drosophila*, one of the best known of these homeotic transformation muta-tions is the Antennapedia gain-of-function mutation. If expression of the gene expands into the cells of the head region, this mutation results in the transformation of fly antennae into legs. These unfortunately affected fruit flies have legs growing out of their heads where the antennae would usually be positioned (see Figure B).

Not just flies . . .

Why are Hox genes such a powerful and conserved mech-anism for directing regional patterning? Perhaps because

Figure A Normal functioning of Antennapedia results in a fly with standard development of the thorax and second leg.

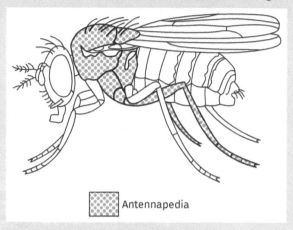

Antennapedia

PhiLiP/Wikimedia Commons

this system allows for a great degree of evolutionary flexibil-ity. Imagine, for example, the differences between a mouse, a snake, and a chicken (Figure C). These three vertebrates have very different-looking body forms. One difference is the number of vertebrae, which we will discuss in the next section. Another difference is the way in which these ver-tebrae are patterned. The chicken has a much longer neck (more cervical vertebrae) than the mouse, while the mouse

Figure B The impact of a mutation in the Hox gene Antennapedia can be clearly seen in these close-up images of the head of a normal fly (left), and one affected by the homeotic transformation (right).

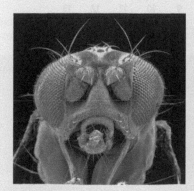

EYE OF SCIENCE/SCIENCE PHOTO LIBRARY

In addition, by studying diverse animals, we can also better understand conserved cellular and molecular mechanisms that have implications for human health.

How do snakes have faces, but worms do not?

In addition to becoming multicellular and developing mechanisms for regional patterning and body segmentation, one clade of animals went further and evolved the host of specialized features that characterizes them as **vertebrates**. The vertebrates are a diverse group, ranging from fish to frogs, crocodiles to hummingbirds and from duck-billed platypuses to people, but they have a surprising number of features in common.

Here we will focus on just one remarkable vertebrate innovation—the sub-population of cells known as the **neural crest**. These cells are only found in vertebrates and are a by-product formed when the neural tube closes as it forms to make a hollow tube. The neural crest cells initially cluster at the most dorsal aspect of the developing neural tube, directly underneath the skin (see Figure 1.11).

Although invertebrates use other cell types to perform some of the same tasks that the vertebrate neural crest cells accomplish, the list is much less comprehensive and the ability to form a face is greatly reduced—which answers our question!

What makes neural crest cells so remarkable is their capacity to migrate to a variety of different locations throughout the developing vertebrate embryo. Once in place, they contribute to a staggering array of structures. For example,

Figure 1.11 The formation of the neural tube and neural crest cells in a vertebrate embryo.

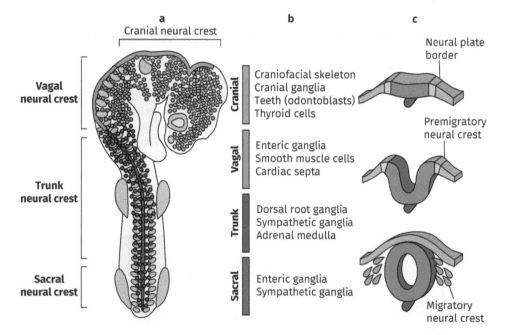

Rothstein, M., Bhattacharya, D., and Simoes-Costa, M. (2018). 'The molecular basis of neural crest axial identity'. *Developmental Biology* Volume 444, Supplement 1, S170–S180

Figure 1.12 The face of a snake and the front end of a worm—now you know why they are so different!

(a)

(b)

(a): Anthony Short; (b): Rhetos/Wikimedia Commons

neural crest cells at the back of the neck migrate into the developing vertebrate head, where they are responsible for generating all the bones and cartilage associated with the face. Without these cells, you wouldn't have a jaw, or a nose—and neither would any other vertebrate (such as the snake shown in Figure 1.12). There would be no man-crushing shark jaws, no funny-shaped duck bills, no elongated horse faces or long, almost comical anteater snouts. Evolutionary biologists think that the contribution of the neural crest to the evolution of the highly functional vertebrate jaw has been critical for the evolution of predation in vertebrates. And what's more, it is this change from filter feeding to active predation and eating more energy-rich food which is seen as key to the continued successful evolution and development of the vertebrates. The possibility that the neural crest cells, and the jaws which evolved as a result of them, led to predation and so to further evolution is a key part of a set of ideas known as the **new head hypothesis**.

It doesn't end there. Neural crest cells are also involved in the development of the functionally divided double circulatory system which is so important to the success of birds and mammals. Without a divided circulatory system, deoxygenated blood cannot be effectively circulated through the lungs to become re-oxygenated. Once this double circulation evolved, the potential for birds to fly, and for mammals to run, jump, swim at great depths, fly, and generally exploit an enormous range of niches became a reality. You can find out more about these evolutionary processes in *Evolution*, also in the Oxford Biology Primers series.

Variations in the neural crest

As you've seen, neural crest cells contribute to many different types of cells and tissues in the body through a series of complex interactions. Variations in these processes can lead to a number of diseases in humans that are known as **neurocristopathies**. Understanding how neural crest cells behave during embryo development can help us to understand these diseases better. In addition, neural crest cells are multipotent stem cells—in other words, they can form a

migrate inappropriately. In the next chapters we will explore the ways in which each of these cellular and molecular processes are used.

Despite the vast diversity of mature body plans (think of the differences between a sea urchin, an annelid worm, a fly, a snake, and a mouse (or look at Figure 1.7)), strikingly similar cellular and molecular mechanisms often underpin the construction process. Therefore, even without a specific interest in evolutionary biology, by studying the morphological, cellular, and molecular strategies that occur across a diverse array of animals, we can understand more about the importance of these strategies. We can also begin to recognize the implications of these strategies for our understanding of other aspects of biology and human health.

Chapter summary

- Developmental biology is a multidisciplinary field that encompasses areas of cell biology, genetics, molecular biology, stem cell biology, evolution, and many more.
- The study of developmental biology is critical for our understanding of human health and disease pathology.
- Evolutionarily, changes in body form (phenotype) are driven by innovations that occur during embryo development.
- Common themes in embryo development and evolution include multicellularity, germ layers, regional patterning, and body segmentation.
- One remarkable example of an embryonic innovation that has led to huge evolutionary advancements is the neural crest—a population of cells responsible for creating vertebrate-specific features such as faces with jaws and divided circulatory systems. These neural crest cells also have many features in common with cancer, making them an invaluable tool for study.

Further reading

Gilbert, M. J. F., and Barresi, S. F. (2019). *Developmental Biology.* 12th edn, Oxford University Press.

This textbook is an iconic survey of processes and mechanisms in developmental biology.

Wolpert, L., Tickle, C., and Martinez Arias, A. (2019). *Principles of Development.* 6th edn, Oxford University Press.

This textbook provides a great overview of many of the principles of developmental biology that we have introduced here.

'What is Evo Devo?' *NOVA.* (pbs.org/wgbh/nova/article/what-evo-devo/)

This webpage provides a great overview of evo-devo (evolutionary developmental biology).

Discussion questions

1. Any change in animal phenotype must occur through innovations made during embryo development. Think about some animals with diverse body plans and how changes during embryo development over the course of generations of evolution could be used to create those body plans. For example, dolphins are mammals with ancestors that were once terrestrial tetrapods. What innovations during embryo development occurred to transform that terrestrial tetrapod body plan into the marine mammal body plan of the dolphin?

2. Imagine how you might use some of the principles of developmental biology that we discussed in this chapter to create mythical animals that might not follow the rules!

Figure 2.1 Expression of different proteins during the first cell division in this *C. elegans* worm embryo gives the cells a unique identity. Mex-5 (magenta) and PIE-1 (yellow) are different proteins that have been labelled with different fluorescent colours in this young embryo. As you can see, at this very early two-cell stage, there is already segregation of these proteins into one of the two cells. The proteins produced allow the cell on the right to act as a communication centre in early embryo development.

(a)

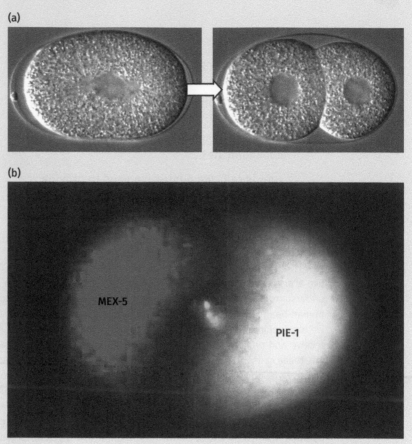

(b)

(a): Photograph courtesy Fabio Piano; (b): Schubert, C. M., Lin, R., de Vries, C. J., Plasterk, R. H. A., and Priess, J. R. (2000). 'MEX-5 and MEX-6 Function to Establish Soma/Germline Asymmetry in Early *C. elegans* Embryos'. *Molecular Cell* 5(4), 671–682

Understanding the language of embryo development

To explore the ways in which cells communicate with one another, we first need to establish language for discussing the huge number of processes that occur during embryo development. These processes are also common in disease pathology and in evolution, so understanding them in the context of embryo development will also serve our broader goals of relating embryo development to other areas of interest.

What's the end-goal of embryo development in animals?

By the end of embryo development, a single cell has been transformed into a complex three-dimensional being. In many animals, including humans, this includes distinguishing the three main **body axes** (head–tail, back–belly, right–left), as well as inside versus outside. This last concept might be a little less familiar, but refers simply to the parts of the body that are maintained inside the animal (for example the heart and digestive tract), compared to those on the outside (for example, skin), as well as all the tissues sandwiched in between (for example, muscles, bones, and connective tissues).

The inside–outside aspects of the animal embryo are called the germ layers, as we mentioned briefly in Chapter 1. Most animals have three germ layers (see Figure 2.2):

- The ectoderm forms the outer layer. In most animals, this is the skin, but—perhaps surprisingly—the brain and spinal cord (the central nervous system) are also formed from this layer. There are also stem cells in this layer, including neural stem cells and stem cells that are used even after embryo development is complete to renew structures in the skin.

- The mesoderm forms the middle layer. In most animals, this includes muscles, bones, and connective tissue (including parts of most of the internal organs). Included here are mesenchymal stem cells, a particularly fascinating type of stem cell found in the adult animal that directs regeneration and repair.

- The endoderm forms the innermost layer. In most animals this includes the digestive tract, the liver, pancreas, and lungs, although all of these organs also have some mesoderm components as well! There are also endodermally derived stem cells critical in the mature organism for renewing cells that line the intestine, as well as cells in the liver.

Figure 2.2 The three basic layers of an animal embryo seen in cross-section. The white area in the centre represents the lumen of a basic gut (surrounded by the endodermal cells that make up the lining of the gut), present in most animals.

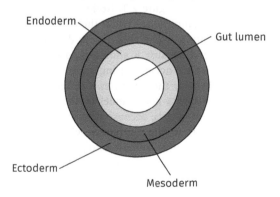

Julia Paxson

Development after embryogenesis?

What happens after embryo development? Post-embryonic development refers to events that occur after formation of the embryo, including:

- **Metamorphosis**—when a radical change in body structure from the immature post-embryonic form (larva or nymph) to produce a sexually mature adult takes place. The strategy of metamorphosis is used in a wide array of animals including some insects, amphibians, and echinoderms such as sea urchins.

- **Post-embryonic growth and patterning**—this involves continuing growth in the mature organism. For example, lobsters continue to grow throughout their life! This is in contrast to organisms such as humans, where growth stops in the mature adult.

- **Tissue homeostasis**—maintaining adult tissues through repair mechanisms which include the use of tissue-specific multipotent adult stem cells (stem cells in the adult that are able to differentiate into several different cell types). This occurs in many animals, including humans.

- **Regeneration**—employing strategies to reconstruct body parts or organs that are prone to damage (from injury or predation). For example, some lizards can regrow their tails if they are lost to a predator, while some worms can regenerate almost their entire bodies. In many organisms, regeneration involves using multipotent stem cells.

- **Ageing**—employing strategies that result in declines in tissue homeostasis, regenerative capacity, and reproductive capacity. Ageing is often related to increased cellular senescence (a specialized form of cell death) and reduced adult stem cell function. In contrast, there is a select group of organisms that exhibit minimal ageing—they have sustained tissue homeostasis, reproductive and regenerative capacities, and live a very long time (for example, animals as diverse as giant tortoises, Greenland sharks, and small planarian flatworms).

Now that we have a framework for understanding how embryo development occurs, we can explore the cellular and molecular mechanisms that are involved. Among these, complex networks of cell communication are a key theme.

Why is cell communication so critical in embryo development, evolution, and disease?

Cell communication is a broad term that refers to all the ways in which cells talk to one another. In a developing multicellular embryo, cell communication is first used to provide positional information that tells each cell where it is relative to the other cells in the embryo, and then starts to provide information about what that cell will eventually become. Depending on the species, this process can occur extremely early (as shown in Case study 2.1), or later in the process of embryo development. However, in all embryos, specification of the germ layers must occur before or during gastrulation.

In addition to the timing of the cell communication events that lead to specification of cell identity, we can also think about whether these events are irreversible. Once specified, can a cell never have a different identity,

Case study 2.1
How worm cells talk in the early embryo

One great example of cell communication in early multicellular embryos has been worked out by scientists in the worm *Caenorhabditis elegans* (*C. elegans*). This worm has very clear signalling mechanisms in the early embryo that have been easy to visualize because the embryo develops outside the parent and is transparent. For these reasons (among many others), *C. elegans* has become a powerful tool in understanding how cells communicate with each other. Remarkably, the same cell communication mechanisms that exist in the worm are also used in many other embryos, including humans.

At the four-cell stage of *C. elegans* embryo development there are two different cell communication mechanisms in play. In this situation, one of the four cells (called P2) is capable of providing positional information to the other three cells about their relative locations using these two different cell communication mechanisms as illustrated in Figure A:

- P2 makes use of the direct contact it has with the neighbouring cell (called ABp) to send it information using the **Notch/Delta signalling pathway** that labels it as the most dorsal cell in the embryo.

- P2 makes use of a common cell signalling pathway with another cell (called EMS) to send it information

using Wnt signalling, labelling it as the most ventral cell in the embryo.

- The third cell (called ABa) does not receive any information from P2, and the lack of signals provides it with information about its position at the head of the developing embryo.

What are the Notch/Delta and Wnt signalling pathways? Read on to understand how these signalling pathways work at a cellular level. The Notch/Delta signalling pathway is named for the mutant fly phenotypes that are generated when the genes are mutated. For example, when mutated, the Notch gene produces flies that have a notch in their wings. Similarly, the Wnt signalling pathway is named for the fact that flies with mutations in the original member of the Wnt family lacked wings (the mutation was called wingless, but was modified to Wnt in vertebrates). Figure A illustrates how both of them work in the four-cell *C. elegans* embryo. At this stage, each cell in the embryo is given a unique identifier—the most anterior cell is known as ABa (AB anterior), the most dorsal cell is ABp (AB posterior), the most posterior cell is P2 (one cell which will form sex cells is always notated as a P cell), while the final cell is EMS (its progeny will form endoderm and mesoderm). The Notch/Delta signalling pathway requires that

Figure A The four cells present in this stage of *C. elegans* development are known as ABa, ABp, EMS and P2. Each cell has a different identity based on positional information that it receives relative to the P2 cell. This cell uses two different mechanisms of cell communication, Notch/Delta signalling and Wnt signalling (including the Wnt receptor Frizzled) and requires that both cells have complementary signalling molecules present.

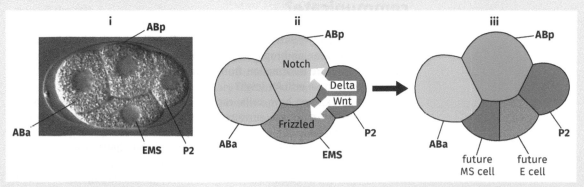

cells compared to its neighbours. For example, in the early sea urchin embryo, transcription factor-induced activation of genetic networks alters the germ layer identity of specific groups of cells, as shown in Figure 2.6. The presence or absence of a single transcription factor (Pmar1) triggers the downstream cascade of activated transcription factors that leads to the specification of cells.

Figure 2.5 On the left side of this illustration, RNA polymerase can only initiate transcription of gene A because the correct combination of transcription factors and activators are present. In contrast, gene B is not transcribed because there is a repressor present, which changes the binding of some transcription factors and therefore also alters binding of RNA polymerase.

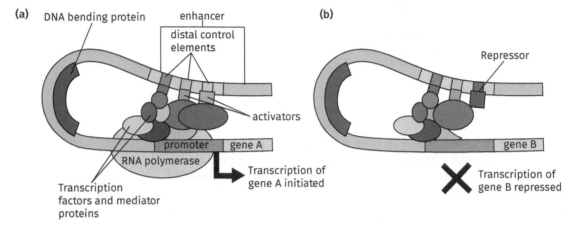

Case study 2.2
Wingless fruit flies and cell communication

One of the classic examples of cell communication in embryo biology is the Wnt signalling pathway. What exactly is this? This particular signalling pathway was first discovered in fruit flies. When components of this signalling pathway are mutated, they give rise to adult flies that are missing wings (as shown in Figure A)—so that pathway was named Wingless (abbreviated to Wg).

This same signalling pathway is present in every other animal in which scientists have looked. The first Wingless homologue was discovered in mice, where it was first known as Integration 1 (Int1). Obviously, mice don't have wings—but when scientists realized that Integration 1 was actually the same gene as Wingless in fruit flies, they created a portmanteau word from Wingless (Wg) and Integration1 (Int1) to form the now universally used

Wnt (Wingless-related integration site) to describe the pathway.

There are many components that together form the Wnt pathway. The **effector protein** (the real workhorse of the Wnt signalling pathway), is a transcription factor known as β-catenin. When no Wnt signal is present, β-catenin is captured by a protein complex in the cell cytoplasm formed by Auxin and GSK3β and undergoes ubiquitination and degradation so that it cannot enter the nucleus. When the Wnt signal is released from neighbouring cells it binds to several different transmembrane receptors at the cell membrane, the most well-known of which are the Frizzled receptor and the LRP receptor. When the Wnt ligand is bound to these receptors, they activate an intracellular protein known as Disheveled that

Figure A The photo on the left shows the result of mutating one of the components in the Wg/Wnt signalling pathway in fruit flies, which results in loss of a wing. The illustration in the middle and right panels shows how the Wnt signalling pathway functions in the two situations where the Wnt signal is either absent (wing missing) or present (both wings are formed).

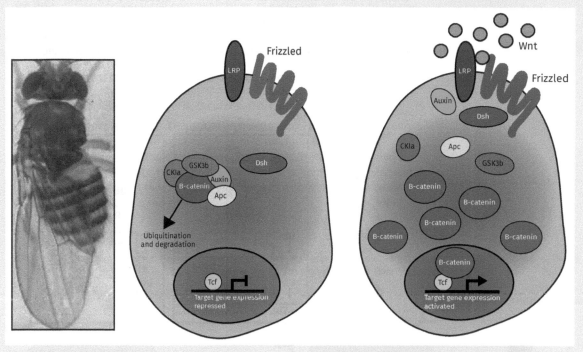

Photograph of fruit fly: Ranganath, H.A., and Tanuja, M.T. (2000). 'Teaching and learning genetics with *Drosophila* 4. Pattern of inheritance of characters when there is interaction of genes or linkage of genes'. *Reson* 5, 59–70. https://doi. org/10.1007/BF02867248; illustrations: Julia Paxson

interacts with the Auxin/GSK3β complex of proteins. This causes the dissociation of β-catenin, which is now able to enter the nucleus of the cell where it acts as a transcription factor to promote the transcription of a host of target genes. So, in this pathway, activation occurs by inhibiting the inhibitor (Dsh inhibiting the inhibiting complex of Auxin/GSK3/APC/CKIalpha). This double inhibition strategy is common in embryo development and enables a high degree of finesse in tweaking both activation and repression of the signalling pathway.

Wnt signalling is used in many different ways in embryo biology, including as a **morphogen** to provide graded positional information, which we will discuss shortly. However, the reach of Wnt signalling is far beyond embryo biology. Wnt signalling is necessary in stem cells

and for regeneration, but inappropriate activation of Wnt signalling is a major cause of colorectal cancer, among others.

❓ Pause for thought

Careful regulation of powerful signalling pathways such as Wnt is critical during embryo development and in mature animals. See if you can imagine ways in which both activation and inhibition of the pathway might occur in Wnt signalling. How might you use activators and inhibitors experimentally to provide loss-of-function and gain-of-function evidence in support of Wnt signalling activity in a developmental process?

- **Paracrine signalling.** This mechanism of cell communication is the one we have already described, where cells receive signals from nearby neighbouring cells. Wnt signalling is an example of paracrine signalling, but there are several other common paracrine signalling pathways in embryo biology, including Hedgehog signalling (another name derived from a mutation found in fruit flies!) and TgfB (transforming growth factor B).

In addition to these pathways, endocrine signalling is a very important category of signalling pathways. Endocrine signalling is a long-distance cell communication tool and it involves transporting the signal (usually a hormone) via some sort of convective flow system (commonly the bloodstream). However, endocrine signalling is not seen in early embryo development because there is not yet any bulk convective flow system!

It's also important to recognize that none of these types of cell communications are used in a vacuum. In fact, cells are receiving hundreds of different inputs, many of which can interact with one another to produce even more subtle cellular responses. For example, four of the more common cell signalling pathways present during embryo development are Wnt, Notch, Hedgehog, and TgfB. Alone, each of these pathways can alter patterns of differential gene expression within the responder cell. But more than that, they can also interact with one another at several different levels, which further alters the final cellular responses. This is illustrated (with appropriate complexity!) in Figure 2.8. Notice that there are multiple feedback loops in which proteins altered by the presence of one pathway can affect activation or repression of other pathways.

How is cell communication used to provide positional information?

A developing understanding of how cells communicate allows us to explore the ways in which cell communication is used in embryo biology. One of the most powerful uses of communication in embryos is in providing relative positional information to cells. But what does this actually mean?

Paracrine signalling pathways can form morphogen gradients

Paracrine signalling pathways are a very flexible way of allowing cells to communicate with one another. For example, it is possible to have a single inducer cell that can disseminate signals to multiple responder cells. Moreover, sometimes responder cells will respond differently depending on how much signal they receive.

If an inducer cell is producing a lot of signal, then some of the signal will diffuse further away from the inducer cell, but most will remain close, forming a concentration gradient. As a result, responder cells that are close by will be exposed to a high concentration of the signal, while responder cells that are further away will be exposed to a lower concentration of the signal.

In embryo development, the way in which responder cells react to paracrine signals is often to adopt a particular cell fate (final identity). In particular, responder cells that are exposed to different concentrations of a paracrine signal along a concentration gradient will often adopt different fates. Even though the

Figure 2.8 This diagram illustrates the complexities of how different cell signalling pathways can interact in the responding cell. Only four different signalling pathways are shown here, including Wnt, Notch, Hedgehog, and TgfB. The reality is even more complicated than this.

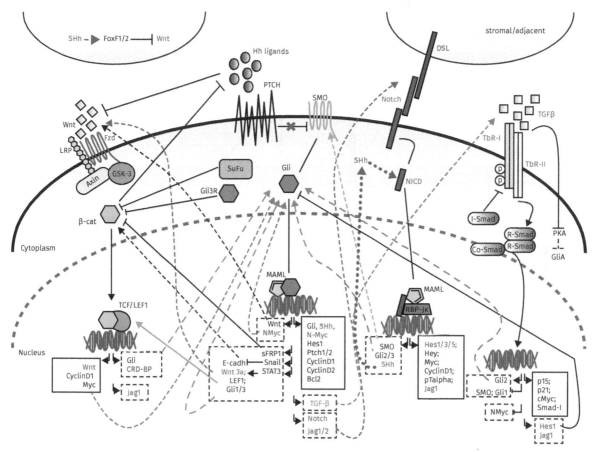

© 2019 Pelullo, M., Zema, S., Nardozza, F., Checquolo, S., Screpanti, I., and Bellavia, D. 'Wnt, Notch, and TGF-β Pathways Impinge on Hedgehog Signaling Complexity: An Open Window on Cancer'. *Front Genet.* 21(10), 711. https://doi .org/10.3389/fgene.2019.00711 This is an open-access article distributed under the terms of the Creative Commons Attribution License (CC BY)

concentration gradient is smooth, often the cells will adopt one of several fates based on receiving signals within particular thresholds. For example, as shown in Figure 2.9, among a population of identical responder cells, high concentrations of the signal will induce the responder cells to adopt cell fate A, an intermediate concentration will induce them to adopt cell fate B, and a low concentration will induce them to adopt cell fate C. In developmental biology, such concentration gradients are known as **morphogen gradients**, because they can induce different cell fates depending on the concentration of signal to which the cell is exposed.

There are many examples of this at different points in embryo development. In particular, morphogen gradients can be used to direct germ layer specification and body axes in the early embryo as well as directing regional patterning along a particular axis in later development. We will see an example of morphogen gradients in regional patterning when we explore limb development in Chapter 5.

Figure B Whole mount fluorescent *in situ* hybridization takes advantage of the fact that individual mRNA molecules can be hybridized to synthetically constructed RNA probes with fluorescent tags attached. When embryos are exposed to RNA probes for two different RNA species (Ets1 and Foxa), only those cells that have the correct mRNA molecules can be visualized. In this case, the RNA probe for Ets1 was constructed with a green fluorescent tag, while the RNA probe for Foxa was constructed with a pink fluorescent tag.

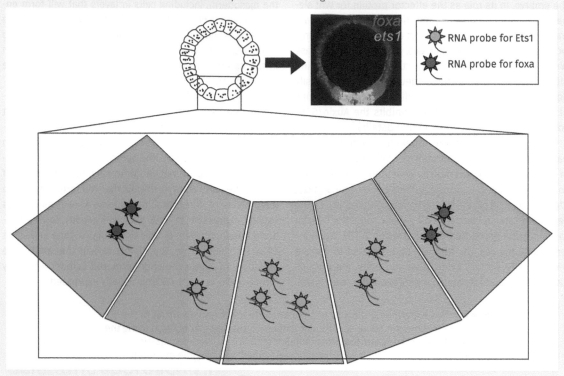

Adapted with permission from *Development*. McCauley, B. S., Akyar, E. Saad, H. R., and Hinman, V. F. (2015). 'Dose-dependent nuclear β-catenin response segregates endomesoderm along the sea star primary axis'. *Development* 142(1): 207–217. doi: https://doi.org/10.1242/dev.113043. Copyright © 2015. Published by The Company of Biologists Ltd

How did the authors follow the activation of these genes in these specific geographic locations within the early embryo? They used a technique called FISH (**fluorescent *in situ* hybridization**). This technique takes advantage of the fact that RNA strands can be experimentally induced to hybridize to another RNA molecule. If the researchers use a small artificially created RNA molecule of the gene of interest and attach to it a fluorescent tag, then any cell with RNA for the gene of interest will fluoresce when observed using a microscope (See Figure B).

What happens in embryos that are bathed in increasing concentrations of LiCl? Can you imagine several different outcomes that might be possible depending on the role of Wnt signalling in early cell fate decisions in the sea star embryo?

In this research article, Brenna McCauley *et al.* show that the higher concentrations of LiCl caused more cells

to show nuclear localization of β-catenin, and importantly caused those same cells to adopt new cell fates as a consequence. This is elegantly demonstrated in Figure C. For example, cells that in the unmanipulated embryo would be specified as endoderm (visualized as expressing the endoderm-specific gene Foxa) will be specified as mesoderm (visualized by expressing the mesoderm-specific gene Ets1) when exposed to increasing concentrations of LiCl.

What's the takeaway?

The authors of this research use a simple approach to generate dose-dependent increases in active Wnt signalling. Their study shows that Wnt signalling acts as a morphogen gradient in the early sea star embryo, specifying different cell fates.

Figure C This figure demonstrates the consequences of exposing sea star embryos to increasing concentrations of LiCl. Increasing numbers of cells show nuclear accumulation of β-catenin and importantly also adopt a different cell fate (for example mesoderm instead of endoderm).

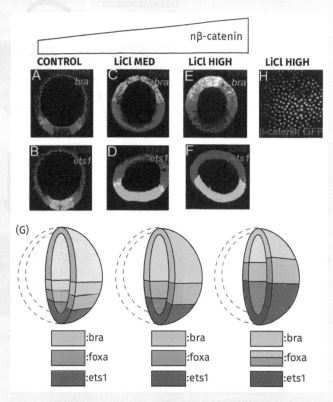

Adapted with permission from *Development*. McCauley, B. S., Akyar, E. Saad, H. R., and Hinman, V. F. (2015). 'Dose-dependent nuclear β-catenin response segregates endomesoderm along the sea star primary axis'. *Development* 142(1): 207–217. doi: https://doi.org/10.1242/dev.113043. Copyright © 2015. Published by The Company of Biologists Ltd

🅰 Pause for thought

In this study, the authors used FISH to visualize the geographic distribution of specific mRNAs. Why would looking for the geographic distribution of the associated protein product be even better? Why might that be impractical? (You'll need to research how proteins are visualized *in situ* to understand.)

Embryonic organizers

The important groups of inducer cells in the early embryo are known as **embryonic organizers**. These cells use cell communication through juxtacrine and paracrine signalling to specify cell fate in surrounding responder cells (see Figure 2.9). In particular, if the embryonic organizer cells are moved to a different part of the embryo, they will continue to produce the same effects among that

Further reading

Gilbert, M. J. F., and Barresi, S. F. (2019). *Developmental Biology*. 12th edn, Oxford University Press.

Wolpert, L., Tickle, C., and Martinez Arias, A. (2019). *Principles of Development*. 6th edn, Oxford University Press.

FlyBase (flybase.org)

This website of *Drosophila* genes and genomics is useful in studying developmental biology.

WormBook (wormbook.org)

This website contains a wealth of information about *C. elegans* and other related worms, including information on their development and many of the signalling pathways that have been elucidated in these animals.

 Discussion questions

1. In this chapter, we briefly mentioned that most of the time, cell communication in embryos results in irreversible changes in cell fate. However, sometimes these changes can be reversed. An example of this is when animal embryos split to form twins. Can you imagine what changes must occur in cell communication for this to happen? How might you explain the fact that many embryos cannot be split to form twins (such as the worm *C. elegans*), while others can (such as humans). To take this even further, how can some mature organisms be split and completely regenerate (such as planarian worms), while most of us cannot?

2. In this chapter, we have seen many incidences where the same signalling pathways are used across many different animal species. What might this tell us about the ways in which current (extant) animal species evolved from common ancestors? How might we use an analysis of specific signalling mechanisms to show that a particular mechanism is not likely to have been conserved across evolution?

3 THE SECRET LIVES OF STEM CELLS

Without question, stem cells are remarkable. Among them, the most remarkable stem cell of all is arguably the fertilized egg, which can give rise to an entire complex three-dimensional organism with hundreds of different types of specialized cells. But the fertilized egg is by no means the only stem cell which has been identified, a fact that is further complicated because defining and identifying stem cells can be tricky and controversial. For example, defining stem cells within the developing and adult animal is complicated by the fact that these cells are constantly regulated by their surroundings and therefore may not act in the same manner that well-characterized stem cells act in laboratory cultures.

Throughout embryo development, different populations of embryonic stem cells contribute to various aspects of the growing organism. Moreover, some stem cells persist in the mature body after embryo development is complete. These persistent stem cells, known as adult stem cells, have many positive roles in the mature organism, including tissue homeostasis, repair, and regeneration. However, stem cells can also behave badly. It is likely that some forms of cancer are initiated by adult stem cells failing to obey the proper regulatory signals. In addition, in many cases stem cell function decreases as an animal ages, which may contribute to some of the diseases that become more common in older people. Paradoxically, some animals such as planarian worms retain excellent stem cell function and consequently seem to avoid ageing!

Studying stem cell biology provides insight into many aspects of living organisms, from understanding embryo development, to manipulating tissue repair and regeneration, to treating cancer, and defying ageing. Figure 3.1 shows several different types of stem cells including the totipotent fertilized egg, adult stem cells in repair and regeneration, and rogue stem cells forming cancer.

three germ layers (see Chapter 2), as well as being specified along the three body axes. So, for example, a cell may be specified as dorsal ectoderm. Later, these cell fates become even more specific. For example, cells in the dorsal ectoderm become specified first as neural and then as different types of neural cells. How and when these specifications occur depend largely on the specific molecular signals that surround the embryonic cells at any given point in embryo development.

In the embryo and in the mature animal, stem cells can be classified in several different ways. The first way is related to the degree of flexibility or **potency** that the stem cells have in the different types of cell fates that their daughter cells can adopt. Figure 3.3 illustrates this progressive cell specification in the

Figure 3.3 This diagram illustrates the reducing potency of cells during embryonic development, as well as the stages during embryo development where embryonic stem cells are often harvested. Some multipotent stem cells are retained in the mature organism.

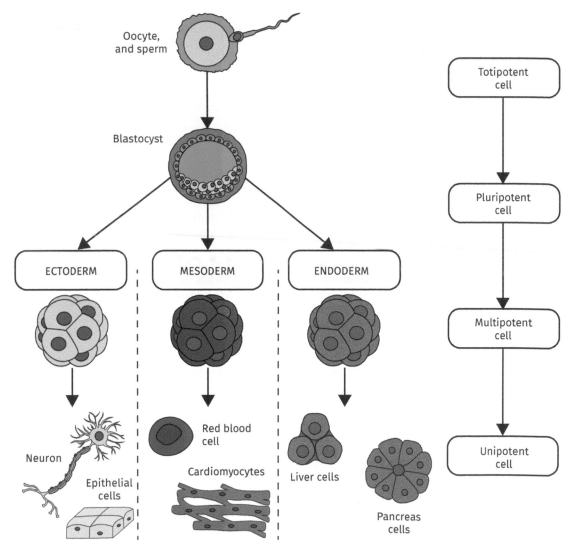

Designua/Shutterstock

Table 3.1 Stem cells can be categorized both by their degree of pluripotency, as well as their place of origin.

Degree of pluripotency	Place of origin
Totipotent stem cells	Zygote
	Early embryo (morula)
Pluripotent stem cells	Pre-gastrulation embryo (inner cell mass)
	Engineered stem cells (iPSCs)
Multipotent stem cells	Post-gastrulation embryo (neural, neural crest, mesenchymal, retained resident stem cells)
	Periparturient (umbilical, placental, amniotic stem cells)
	Mature organism (retained adult stem cells)
	Engineered stem cells

developing embryo (see also Table 3.1 for a summary). Cells that are unipotent are not considered stem cells.

- **Totipotent stem cells**—these are extremely potent stem cells that can differentiate into any cell type found in the embryo, including the placenta and other extra-embryonic tissues. This category really only includes the fertilized egg, and, in the case of humans and other mammals, the cells that are the product of the first few cell divisions. In early human embryo development, the first few cell divisions lead to a ball of **totipotent** stem cells that can each produce an entire embryo (including the placenta) if isolated. Soon after this, the placental cells will become distinguishable and clearly different from a clump of cells known as the inner cell mass that will form the embryo. After this, cells in the inner cell mass are no longer considered totipotent because they cannot make the placenta.

- **Pluripotent stem cells**—these are also extremely potent stem cells that can differentiate into any cell type in the embryo proper (not counting the placenta and other extra-embryonic tissues), including cell types in all three germ layers. This includes many stem cell populations in the very early embryo, as well as stem cells in adult organisms (such as planarian worms) that have the capacity for whole-body regeneration. Finally, some types of engineered stem cells such as induced **pluripotent** stem cells, have this capacity.

- **Multipotent stem cells**—these are slightly less potent stem cells but they still differentiate into several different cell types. These stem cells are often restricted by germ layer—for example, ectodermal **multipotent** stem cells may be restricted to only differentiating into ectodermal cell types. However, there is some plasticity here, since some mesenchymal stem cells can differentiate into neurons, and some neural crest stem cells can differentiate into both mesodermal and ectodermal cell types. Often, multipotent stem cells are found in adult tissues, where they are restricted to differentiating into cell types specific to that particular tissue. These retained stem cells are induced to become quiescent and retain their potency long after embryonic development is complete. These become adult stem cells and are critical for tissue homeostasis, repair, and regeneration in the mature animal.

Figure 3.4 Neural crest stem cells originate at the dorsal aspect of the closing neural tube (shown in relation to the cross-section across the developing embryo on the left). They migrate to a variety of locations where they generate many different cell types that contribute to a variety of diverse structures. In doing so, many of these neural crest cells retain multipotent status until they adopt their final cell fate after migration.

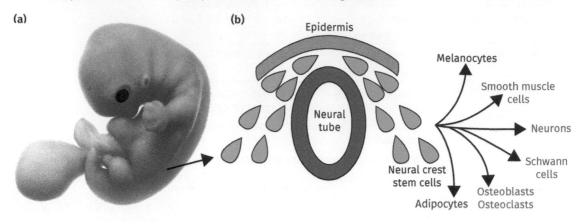

(a): SciePro/Shutterstock; (b): Julia Paxson

Although embryonic cells generally experience a loss of potency as the embryo develops, there is a lot of variation about when, where, and if this loss happens. Some stem cells in the embryo are induced to remain potent for longer than others. For example, neural crest cells, which we mentioned in Chapter 1 in their capacity for generating diverse structures such as the face, retain their potency relatively late so that they can adopt one of several diverse cell fates as illustrated in Figure 3.4. There is recent evidence that some neural crest stem cells may retain their multipotent status in the mature organism. This is currently an area of interest for potential regenerative applications. In addition, there is also evidence that some retained neural crest stem cells may contribute to certain cancers in adult mammals.

This categorization strategy for stem cells sometimes gives the misleading impression that loss of potency and commitment to cell fate is irreversible, but this is not always true. There are both experimental and naturally occurring examples of cells which have differentiated to a particular cell fate, but have then been induced to reverse course and revert to a more potent undifferentiated stem cell. The best-known example is a type of engineered stem cell known as **induced pluripotent stem cells (iPSCs)**, which we will discuss later in this chapter.

The second way that stem cells can be classified is based on where they are found:

- **Embryonic stem cells**—these are stem cells directly related to embryo development, so they are part of the very early embryo. Since traditionally these cells are harvested by killing the embryo, the use of *human embryonic stem cells* has been controversial. However, there are other ways of harvesting very early human embryonic stem cells that do not destroy the embryo. Moreover, many other types of stem cells are present during later embryo development. For example, neural stem cells, neural crest stem cells, and mesenchymal stem cells.

- **Adult stem cells**—these are stem cells which are still present at the end of embryonic development. It turns out that there are a large number of different types of stem cells that exist beyond embryo development, which we will discuss in more detail. However, in mammals this category includes stem cells associated with the mature foetus, such as stem cells associated with the placenta, the amniotic sac, amniotic fluid, and the umbilical cord.

- **Engineered stem cells**—most commonly referred to as induced pluripotent stem cells. In addition to naturally occurring stem cells, it is also possible to form **engineered stem cells** from other types of cells. Indeed, it's possible to take skin cells and, by giving them the right combination of signals, convince them to become stem cells. Inducing differentiated cells to become undifferentiated was once thought impossible, but we now know it can be done.

The changing roles of stem cells across a lifetime

As embryonic development progresses, the dividing stem cells gain more information about their position and their eventual cell fates, which we will discuss in more detail in Chapter 4. However, even as cells in different tissues are undergoing differentiation, some cells are retained as multipotent stem cells. These multipotent stem cells have one of two broad functions in the mature organism—tissue homeostasis and tissue repair, including tissue regeneration:

- **Tissue homeostasis.** There are a variety of different tissues in the mature organism that need to be replaced on a regular basis. In humans, epithelial surfaces such as the skin and the intestinal lining are constantly shed and need to be replenished with new cells. For example, the epithelial cells that line the inside of the intestinal tract die and are shed roughly every 3–5 days, so there has to be a consistent source of new intestinal epithelial cells to replace them. These new cells arise from intestinal stem cells that live in the valleys of the folds of the intestinal tract (known as the crypts). Daughter cells of the intestinal stem cells slowly migrate up towards the top of the folds, known as the villi, where they differentiate into functional intestinal epithelial cells, and then are rapidly shed (see Figure 3.5). Intestinal stem cells are generated during the formation of the intestinal epithelial lining in the embryo. Communication between cells ensures that these stem cells are retained in their undifferentiated form, rather than becoming other types of differentiated intestinal epithelial cells.

- **Tissue repair.** Many tissues in the mature organism have a small scattered population of multipotent stem cells that are generally quiescent (not dividing), unless the tissue becomes damaged. For example, muscle stem cells, also known as satellite cells, perch on the outside of complex multi-nucleated muscle fibres. They are dormant—not dividing or differentiating. However, they are activated by signals associated with inflammation and damage. When the muscle stem cells receive these signals, they respond by re-entering the cell cycle and dividing rapidly to produce new progeny which are used to repair the damaged muscle fibres. Once the damage is repaired, the muscle stem cells exit the cell cycle and re-enter the dormant quiescent state. During muscle development in the embryo, muscle progenitor cells (known as myoblasts) become differentiated to form the specialized fused multinucleated myotubes that make

Figure 3.5 (a) A cross-section of intestines. (b) Intestinal stem cells are located in the crypts (magenta), shielded from differentiation during embryonic development of the intestinal epithelium. (c) These intestinal stem cells are very active, dividing to provide new cells which are pushed up towards the villi, where they differentiate into one of several different types of intestinal epithelial cells before relatively quickly undergoing cell death and being sloughed off into the intestinal lumen.

(a)   **(b)**

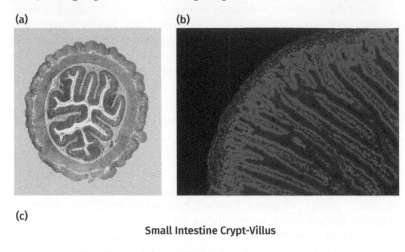

(c)

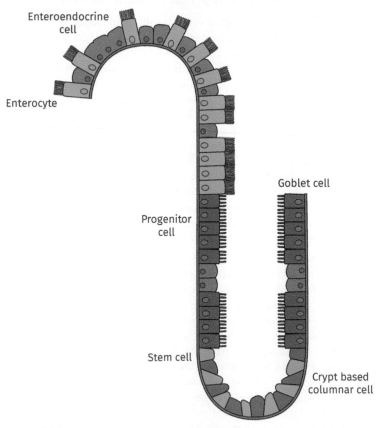

(a): Jubal Harshaw/Shutterstock; (b): Virginie Thomas/Shutterstock; (c): joshya/Shutterstock

up the muscle fibres of muscle tissue. However, during this process, some myoblasts are held in suspension so that they do not differentiate, using a process known as juxtacrine Notch signalling. These will eventually become muscle stem cells and will not become part of the fused myofibre. These muscle stem cells can be used to repair or grow muscles, either after injury or in response to exercise, as illustrated in Figure 3.6.

- **Tissue regeneration.** Depending on the animal, tissue repair may be limited to small areas within an already existing tissue, or it may be much more extensive. For example, many animals are capable of regenerating entire organs or appendages. Salamanders can regenerate entire limbs, tails, hearts, and even eyes. However, the masters of tissue regeneration are invertebrates, which can regenerate entire segments of their bodies. For example, sea stars can regenerate their bodies starting with only a limb. Some worms (planarian worms and acoels for example) can regenerate most of their bodies (inducing their brains!) from small segments of tissue (Figure 3.7). We will explore this fascinating area of biology (and why we are relatively poor regenerators!) later in the book.

The changes in the behaviour of these organ-specific **adult stem cells** are largely controlled by their surrounding environment, the stem cell niche. This niche is made up of cells and extracellular materials working together to regulate stem cell proliferation (division) and differentiation. For example, in the niche surrounding intestinal stem cells, there are support cells that provide paracrine signals such as Wnt to regulate how much the stem cells divide. In addition, changes in the tension and structure of the basement membrane and surrounding extracellular matrix also influence division characteristics in these stem cells (see Figure 3.8).

Communication between the stem cells and surrounding cells in the stem cell niche is critical for the appropriate regulation of stem cell proliferation and differentiation of resulting daughter cells. If signalling from the niche changes, or if the stem cells alter their ability to perceive and respond to these signals, the finely tuned regulation of these powerful cells can unravel. One potential consequence of dysregulation of stem cell proliferation and differentiation is

Figure 3.6 In the mature organism, either during exercise or after injury, muscle stem cells (also known as satellite cells) are used to repair skeletal muscle tissue. These satellite cells remain quiescent until activated, when they proliferate (divide to create more stem cells) and differentiate into new muscle progenitor cells (myoblasts), which fuse together to form the mature myofibre.

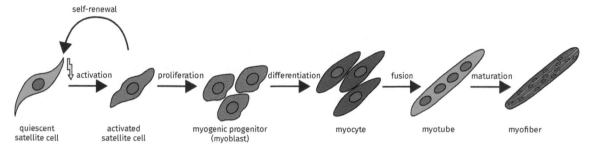

Schmidt, M., Schüler, S.C., Hüttner, S.S. *et al.* (2019). 'Adult stem cells at work: regenerating skeletal muscle'. *Cell. Mol. Life Sci.* 76, 2559–2570

the amount of DNA damage, **telomere** length, and external signals. In many cancers, the proteins involved in cell cycle arrest are mutated, including both **tumour suppressor genes** such as P53 that usually arrest the cell cycle, as well as proto-oncogenes such as Ras and Myc that can block these cell cycle arrest points. When mutated, proto-oncogenes are named **oncogenes** for their capacity to drive the high proliferation rates seen in cancerous cells. However, there are other stem-like properties commonly seen in cancerous cells that cannot be accounted for by these mutations. Therefore, it is possible that many types of cancer are formed from mutations that occur in stem cells or progenitor cells. These possible stem cell origins may have implications both in the properties of cancers to undergo metastasis, as well as to resist the effects of chemotherapy, as illustrated in Figure A and as we will discuss.

* **Cancer metastasis.** Many forms of cancer are particularly vicious because of their ability to metastasize

Figure A Cancer stem cells many be formed by multiple mutations, especially in cell cycle regulation genes in stem cells or highly proliferative progenitor cells. This theory accounts for many of the stem-like properties seen in cancer. Furthermore, the presence of cancer stem cells in a tumour may also account for the capacity of the tumour to metastasize to distant locations in the body.

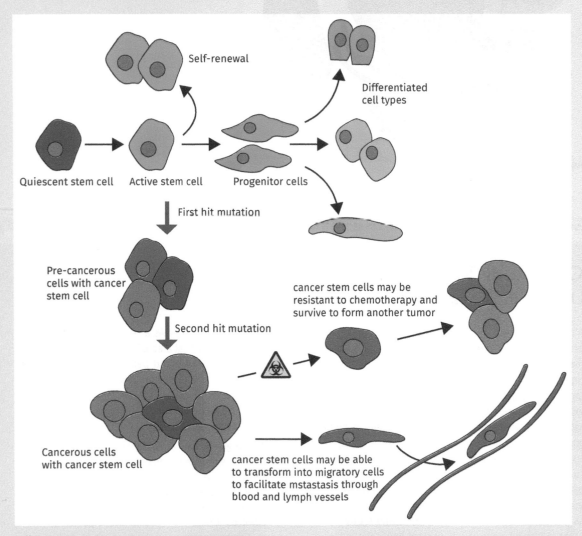

Julia Paxson

and spread to different sites across the body, where they can form additional reservoirs of cancer cells. This capacity is largely reliant on the ability of cancer cells to migrate away from their site of origin, as illustrated in Figure B. For tumours with epithelial origins, this involves undergoing epithelial-to-mesenchymal transition (EMT), a phenomenon that occurs at several points during embryo development as cells change position relative to one another (see Chapter 1 for more details about the process of EMT in embryo biology). The suite of proteins needed to undergo EMT is also commonly found in stem cells. Therefore, understanding more about the process of EMT and how it can be triggered in stem cells is also critical to our understanding of how to limit this process in epithelial cancers.

- **Resistance to chemotherapy.** If cancer stem cells are present in a tumour, driving tumour expansion and metastasis, then these cells must be eliminated for the therapy to be effective. Traditional chemotherapeutic agents are engineered to attack rapidly dividing cells, which often includes cancerous cells.

Figure B Cancer cells in epithelial cancers such as carcinomas can undergo EMT, where they lose polarity, lose attachments to the basement membrane, and become mobile. This enables them to migrate away from the site of the original tumour, invade the circulatory system, migrate to distant sites where they then move back out of the blood vessels, undergo MET (the reverse of EMT), and potentially colonize other tissues to form new reservoirs of cancer cells.

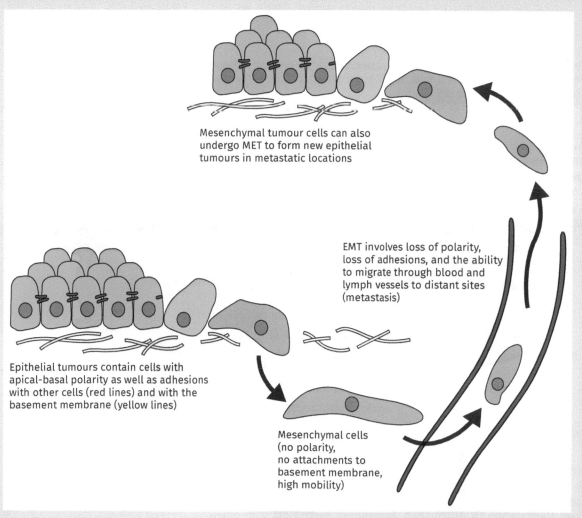

Mesenchymal tumour cells can also undergo MET to form new epithelial tumours in metastatic locations

EMT involves loss of polarity, loss of adhesions, and the ability to migrate through blood and lymph vessels to distant sites (metastasis)

Epithelial tumours contain cells with apical-basal polarity as well as adhesions with other cells (red lines) and with the basement membrane (yellow lines)

Mesenchymal cells (no polarity, no attachments to basement membrane, high mobility)

Julia Paxson

types derived from the same germ layer. On the other hand, sometimes stem cells may be too potent. For example, using totipotent embryonic stem cells for tissue repair or regeneration might risk those cells differentiating into the wrong cell types, or becoming dysregulated and forming cancer instead. There are some documented examples of this happening—for example the formation of space-occupying tumours after the experimental treatment of spinal cord injuries with highly potent stem cells.

- **Mode of action.** Stem cells can work in a variety of ways. Perhaps most intuitively, stem cells can work by integrating themselves into damaged or diseased tissues where they create new tissue. However, this is not the only (or even perhaps the most common) mode of action. Stem cells can also work by releasing paracrine signals that stimulate other resident cells to help with repair and regeneration of damaged or diseased tissues. Finally, some types of mesenchymal stem cells can also participate in immune-modulation, which can be important in repairing damage.

- **Source.** Once the type of stem cell has been selected, perhaps with some idea about the mode of action that it might have, given the intended application, there are still more decisions to make. Among these is the source of the stem cells—will they be derived from the patient (autologous) or come from another donor (allogenic). There are advantages to using cells derived from the patient (whether adult stem cells or induced pluripotent stem cells), including no immune reaction from the recipient and no potential for introducing foreign pathogens. However, the age of the donor may significantly contribute to the functionality of the stem cells—they become much less active as we age so it may be more successful to use stem cells from a younger donor, in spite of the immune challenges. Fortunately, stem cells tend not to provoke the same types of immune reactions as donated organs (because they have very few protein markers on their cell surfaces), so reducing the risk of using allogenic stem cells.

What are the potentially unique uses of engineered stem cells?

There are so many sources of stem cells, why would we want to try and engineer stem cells from a patient's own differentiated cells? Most of the available sources of stem cells are multipotent adult stem cells. These appear to have great potential for therapeutic use, but there are some special applications that require a patient's own cells that have been engineered to generate induced pluripotent stem cells (iPSCs):

- **Autologous stem cell therapy for tissue repair.** As we have already discussed, there are many forms of stem cells that might be useful for this. The potential advantage of iPSCs is that they are derived from the patient, so are not rejected (although that might also be true for other sources of allogenic stem cells such as the mesenchymal stem cells or MSCs that are used in bone marrow transplant).

- **Gene correction.** It is possible that iPSCs could be used in patients with genetic diseases to provide copies of corrected genes that could be locally administered to the relevant tissues—for example, the treatment of cystic fibrosis or sickle cell anaemia.

- **Drug testing.** IPSCs could be used (either as stem cells or as more specific differentiated cells) to test drugs. This is recognized as increasingly important since individuals can react very differently to the same drug. Therefore, the patient's own cells could be used in testing to limit the adverse affects that might occur. This could be important for a wide range of drug applications.

- **Drug development.** Having iPSCs from individual patients represents a potentially huge leap forward in drug development. Human iPSCs can be engineered to produce a large number of different types of differentiated cells. When combined in different ways they produce tiny, 3-D tissue cultures known as organoids, which carry out some of the functions of organs such as the kidneys. These organoids can be used to evaluate drug function and other therapeutic applications.

However, it is important to recognize that iPSCs would not be useful when treatment is needed fast—for example moderating the immune system in acute sepsis. At the moment, it takes weeks or months to correctly engineer, test, and potentially redifferentiate patient-specific iPSCs.

How are stem cells isolated?

While it is possible to study stem cells within the developing organism, it is also often useful to be able to study stem cells in culture. In addition, stem cell culture is necessary for clinical therapies that aim to deliver stem cells to the human or veterinary patient. **Cell culture** is a method of convincing cells to grow outside the body. The natural growth environment is mimicked by growing these cells on sterile plastic Petri dishes with a special growth broth that contains all of the nutrients that the stem cells need to survive.

How are stem cells isolated? It depends on what type of stem cell you want to study. For example, embryonic stem cells were traditionally isolated by breaking open the human embryo after the inner cell mass had formed and growing those pluripotent stem cells in culture. However, in order to do this, the embryo must be destroyed, which has clear ethical implications. More recently, work pioneered in fertility clinics working with IVF (*in vitro* fertilization) embryos has revealed that 4–8 cell human embryos may safely have one or more cells removed without compromising the survival of the embryo. The single (totipotent) stem cells that are removed can then be grown in culture (see Figure 3.9).

For adult stem cells, there are a variety of ways to isolate stem cells from the tissues. Often, resident stem cells can be identified by the specific combinations of proteins that stud the outside of their cell membranes. These specific protein combinations can be used to identify the cells and isolate them. This is used particularly with ectodermal stem cells (such as intestinal epithelial stem cells). The specific functional properties of mesenchymal adult stem cells make them relatively easy to isolate. You might remember that, unlike epithelial cells, mesenchymal cells are migratory and unattached to the basement membrane or to each other. As a consequence, it is possible to place a small piece of tissue (from the lung for example) into a Petri dish, cover it with growth media, and observe the

Figure 3.11 Induced pluripotent stem cells (iPSCs) are constructed using differentiated (somatic) cells from individual patients. Using a remarkably simple cocktail of four growth factors, these somatic cells can be induced to form pluripotent stem cells that can potentially be used in a wide range of applications.

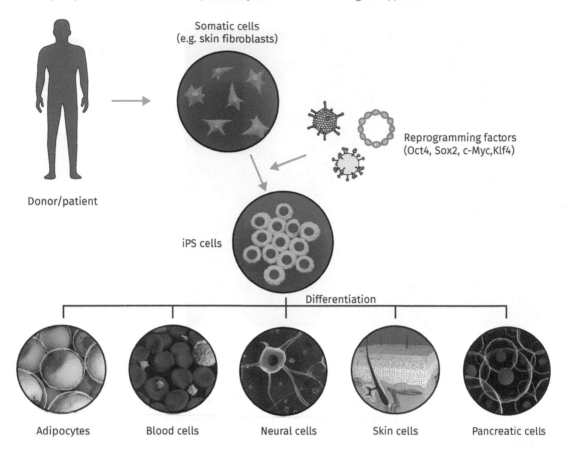

Meletios Verras/Shutterstock

cells to replace bone marrow lost during treatment for leukaemia, there are very few stem cell therapies for human disease that have been convincingly demonstrated to be effective. This is despite a good body of scientific literature showing that these same treatments are effective in rodent models. Why does such a discrepancy exist? There are many possibilities, including the differences between human and rodent physiology, inconsistencies in the ways that the stem cells are isolated and processed, as well as the heterogenous nature of human diseases compared to experimental animal models. It does seem likely that before too long, effective stem cell therapies will be worked out in humans. As this progresses, it will be increasingly incumbent upon us to pay particular attention to the protocols used to isolate and test the stem cells before use– where do they come from, how are they isolated, how are they tested, how are they administered to the patient? What is the timeframe for stem cell isolation and treatment? Where (geographically) are the stem cells isolated and grown? Only by understanding the nuances of stem cell biology will it be possible for us to use them effectively and ethically in research and in clinical therapies.

Figure 3.12 Stem cells are being used increasingly to repair tendon, ligament, and joint problems in horses. The hope for the future is that people will gain the same benefits—but there are both scientific and ethical questions to answer first.

Anthony Short

Scientific approach 3.1
Can stem cells repair spinal cord injury?

Repair of tissues after injury is a major area of investigation in stem cell biology. Within this field, there is a great deal of interest in finding effective treatments for spinal cord injury (SCI). In humans, most spinal cord injuries are created through crushing events that leave the nerves within the spinal cord permanently damaged and unable to repair themselves. Remarkably, it is possible to use multipotent stem cells to repair nerve axons in the laboratory, but translating this into a therapy that is effective in the human spinal cord after a crush injury has been challenging.

As discussed previously, there are many considerations when trying to design an effective therapy. For example, should we use autologous or allogenic stem cells, and would adult stem cells (including mesenchymal stem cells or potentially adult neural stem cells)

- Cancer stem cells may represent a sub-population of abnormally functioning stem cells that have the capacity to migrate, proliferate, and avoid chemotherapeutics. Better understanding these cells may therefore lead to improved cancer treatments.
- In addition to their natural roles in the body, stem cells have many potential applications for research and clinical therapies, including repair of damaged and ageing tissues, a tool to model specific diseases and customize individual patient therapies, a vehicle to deliver medications, and a tool to reduce animal use in research and testing.
- Using stem cells in research and therapies requires careful consideration of which stem cell type is used, its potency, and its potential mode of action, as well as how it is isolated and manipulated.
- As stem cells become more commonly used in clinical therapies, it will be increasingly important that we understand the detailed science behind these remarkable cells.

Further reading

Gilbert, M. J. F., and Barresi, S. F. (2019). *Developmental Biology*. 12th edn, Oxford University Press.

Lee, H., Lee, H. Y., Lee, B. E. et al. (2020). 'Sequentially induced motor neurons from human fibroblasts facilitate locomotor recovery in a rodent spinal cord injury model'. *eLife* 9:e52069. doi: https://doi.org/10.7554/eLife.52069
This is the paper referenced in SA3.1 Figure A.

Wolpert, L., Tickle, C., and Martinez Arias, A. (2019). *Principles of Development*. 6th edn, Oxford University Press.

Zakrzewski, W., Dobrzyński, M., Szymonowicz, M., et al. (2019). 'Stem cells: past, present, and future'. *Stem Cell Research & Therapy* 10, 68. doi: https://doi.org/10.1186/s13287-019-1165-5
This is a recent review of the challenges that must be overcome in order for stem cell therapies to be effective and widely available.

Discussion questions

1. Do your own research! What types of stem cells are currently being tested in clinical trials? What can you tell about their origin, how they were isolated, how they were manipulated, and what the proposed mechanism of action is for their therapeutic use?
2. Are there any commercially available stem cell treatments available in the country where you live? How are these therapies regulated? How might commercial facilities try to work around these regulations to offer possibly unproven stem cell treatments?
3. Which conditions currently have successful stem cell therapies available?

4 EMBRYO ORIGAMI

Origami is the art of paper folding, originally developed in Japan. By making skilful folds, origami experts can turn a flat sheet of paper into beautiful three-dimensional shapes, from birds to flowers to frogs to rhinos (see Figure 4.1). It is almost magical to see it happen before your eyes.

Figure 4.1 Origami cranes—these delicate 3-D birds were made from single flat sheets of patterned paper.

Anthony Short

How do we know? If the four cells are split apart from one another, each one can create an entire sea urchin. However, the cells in the 8-cell embryo are no longer equivalent to one another. If these eight cells are split apart from one another, the posterior four cells form almost normal-looking sea urchins. In contrast, the top four cells (lacking positional information from the organizer) form an undifferentiated mass of cells. Can you imagine how changing the orientation of the cell divisions would alter these outcomes? For example, what if, experimentally, the second cell division in the sea urchin embryo was altered so that the cells in the anterior half are divided from cells in the posterior half where the maternally derived cytoplasmic factors are segregated? Given that information, which cells might you predict would be able to form entire animals if isolated and which cells might not?

What about humans?

What happens in human embryos? During maturation of the egg, many macromolecules certainly accumulate in the egg cytoplasm, but there is not yet evidence that any of these are tethered or segregated to a particular region of the cell. After fertilization, and during the first few rounds of cell division, the embryonic cells appear to remain equivalent to one another (see Figure 4.3). Indeed, it is possible to remove cells or split the embryo into several parts at this stage (known as a morula)—and still form multiple individual embryos. Of course, it is hard to provide strong scientific evidence in human embryos, since there is so much ethical controversy involved in any experimentation using human embryos. Further, not all mammalian embryos behave the same way, so it is not always possible to extrapolate from rodent embryos to human embryos. For example, mouse embryonic stem cells do not appear to retain their totipotent nature for as long as humans. Perhaps this is because mice tend to have multiple embryos in any given pregnancy, so from an evolutionary perspective, developmental flexibility may be less advantageous than it is in human pregnancies.

Figure 4.3 In human embryos, cells in the first few rounds of division appear to be equivalent and totipotent in their capacity to form an entire embryo if isolated. After formation of the blastocyst, a group of cells known as the epiblast will form the embryo itself and at this point, the cells appear to start the process of cell fate specification and become multipotent.

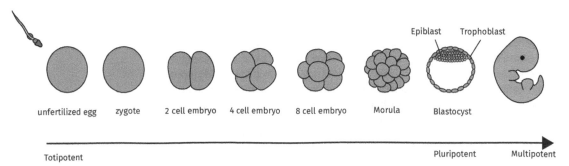

Julia Paxson

In human and other mammalian embryos, cell communication appears to be a more common mechanism for providing positional information, rather than segregated maternally derived cytoplasmic factors, By the late morula and early blastocyst stage, cells use a variety of signalling pathways to tell them whether they are positioned on the inside or the outside of the embryo. This appears to be the point at which these embryonic stem cells lose their totipotent qualities and transition to becoming pluripotent (see Figure 4.3). Internally positioned cells tend to form the **epiblast** (also known as **inner cell mass**). The epiblast then forms the pancake-like embryonic structure which undergoes gastrulation. However, even within the epiblast there may still be considerable flexibility in the cells over their eventual fate in the body. The evidence for this comes from the possible late timeline of monozygotic (identical) twin formation in humans (see Case study 4.1).

Case study 4.1
How and when do human twins form?

Human twins form in two basic ways, although there are many additional, much rarer variations. Dizygotic twins—also known as fraternal twins or non-identical twins—form from the simultaneous fertilization of two eggs by two sperm, so dizygotic twins do not share identical genomic information—they are often different sexes. There are even cases of dizygotic twins with different fathers! On the other hand, monozygotic or identical twins, as shown in Figure A, form from a single fertilized egg. At some point, the developing embryo splits to form two individuals who are (almost) genetically identical. Globally, around 3.5–4 births in every 1000 will be identical twins. The occurrence of dizygotic, non-identical twins, is much more varied, ranging from 8 per 1000 births in much of Asia to 17 per 1000 births in much of Africa. In many developed countries, the incidence of dizygotic twin births has risen with the increased use of modern technologies to overcome infertility.

Growing in the uterus

As a human embryo develops, it also produces the **placenta**, which works with the maternal tissues to provide the food and oxygen needed as the pregnancy progresses, and the **extra-embryonic membranes** which surround the embryo and contain the amniotic fluid. Figure B illustrates how these extra-embryonic membranes are arranged in a single human embryo. The placenta develops from part of the chorion, forming a thickened disc that penetrates into the uterine lining.

What happens with twins? Dizygotic twins are almost always dichorionic and diamniotic. This means that they are each surrounded by their own amniotic and chorionic membranes. You might expect that monozygotic twins would therefore share both their membranes and placenta. In fact, monozygotic twins may be born with their own amniotic and chorionic membranes, or they can share one or both.

How late can human embryos form monozygotic twins? The truth is that we really don't know. Potentially, one way to tell is by examining the arrangements of the extra-embryonic membranes in monozygotic twins. Twins that are monoamniotic and monochorionic are traditionally thought to originate from late splitting of the epiblast, once the early stages of the membranes have already been formed. However, an alternative theory suggests that perhaps twinning embryos always split early—maybe at the two-cell stage—and then their amnions and chorions fuse later. Both of these possibilities are illustrated in Figure C. The question is, how would we tell the difference between these two very different possible scenarios? As yet, no-one has the answer.

we once imagined, and that the fate of cells can be changed or even reversed. This concept is illustrated in Figure 4.4, where the process is modelled as a series of ridges and valleys in the developmental landscape—where the valleys represent distinct cell fates, and the ridges represent the barriers to cells moving between these different cell fates. Cells are more likely to end up in the valleys through the influence of both internal and external factors. However, with enough influence, they can move across the ridges into new valleys, which represent new cell fate identities.

Transdifferentiation is the term used for differentiated cells that transform into another differentiated cell type without passing through a multipotent stage first. **Dedifferentiation** is the process by which differentiated cells transform into a less differentiated (multipotent or pluripotent) stem cell. Both transdifferentiation and dedifferentiation occur naturally, but can also be induced experimentally, which is useful both in research and for developing clinical therapies.

Figure 4.4 In an imagined developmental landscape, the route to a particular cell fate occurs in valleys, with barriers to these cells represented as ridges. Generally, pluripotent (green) cells and multipotent (blue and yellow) cells acquire cell fates through internal and external information. However, it is also possible for differentiated cells to either transdifferentiate to a different cell type (orange arrows) or dedifferentiate to a stem cell with greater potency (blue arrows).

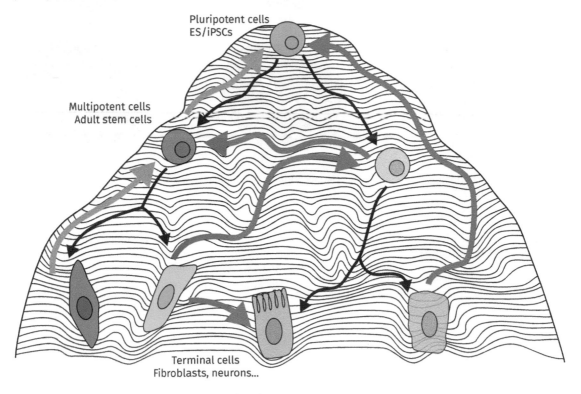

Pluripotent cells
ES/iPSCs

Multipotent cells
Adult stem cells

Terminal cells
Fibroblasts, neurons...

Gam, R., Sung, M., and Prasad Pandurangan, A. (2019). 'Experimental and Computational Approaches to Direct Cell Reprogramming: Recent Advancement and Future Challenges'. *Cells* 8(10), 1189. https://doi.org/10.3390/cells8101189.

Figure 4.5 In mammalian embryos cell fate acquisition is evident at the beginning of gastrulation (known as primitive streak formation), although it may be initiated earlier than this. Pluripotent stem cells are first induced to acquire cell fates associated with one of the three germ layers. After gastrulation is complete, tissue specification begins with more defined cell fate acquisition. This entire process is largely determined by external signals from both neighbouring cells and from the environment.

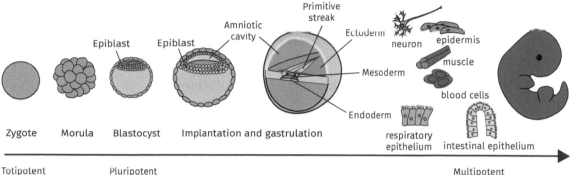

Julia Paxson

How does this process play out in the embryo? Figure 4.5 illustrates the progression of cells as they move in a conventional way though the valleys of the developmental landscape in a mammalian embryo. Cells that are pluripotent in the very early embryo quickly acquire identity as either ectoderm, mesoderm, or endoderm in a process called germ layer induction. The cell movements of gastrulation rearrange these cells relative to one another. Simultaneously, the cells receive further information from each other and from their surroundings, and this mass of new information helps in further determining the fate of the cell. For example, in many animals, Wnt signalling is important in the acquisition of both mesodermal and endodermal cell fates. After gastrulation, further signalling causes mesodermal and endodermal cells to differentiate even more. However, many of these cells are still multipotent and capable of high rates of cell proliferation as embryo development progresses.

At the same time that cells are acquiring progressively more specialized cell fate identities, large-scale movements of cells relative to one another are taking place. The relative positions of cells are controlled in the very early embryo by specifying the orientation of cell division planes. Indeed, in plants this is the major mechanism for controlling cell movements relative to one another. However, in animal embryos, cells can also move either in sheets of epithelial cells, or by acquiring a mobile mesenchymal cell phenotype. The most studied example of cell movements in animal embryos is the process known as gastrulation.

Gastrulation: everybody does it

Gastrulation is a process that helps to generate functionally specialized tissues in a developing organism, by physically rearranging germ layer cells to assume their final positions relative to one another. By the end of gastrulation in **triploblastic** animals, the mass of cells that formed the early embryo is rearranged into a three-dimensional, often elongated, embryo. The endoderm and mesoderm have moved inside the embryo and been repositioned relative to one

Figure 4.7 This illustration shows a cross-section through the pancake-like human embryo during gastrulation. The embryo is suspended between the amnion (seen here in blue) and the yolk sac (seen here in yellow underneath the embryo). As gastrulation progresses, the primitive streak grows from posterior to anterior of the embryo.

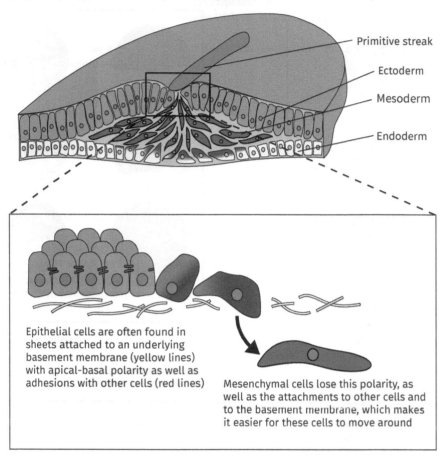

Primitive streak

Ectoderm

Mesoderm

Endoderm

Epithelial cells are often found in sheets attached to an underlying basement membrane (yellow lines) with apical–basal polarity as well as adhesions with other cells (red lines)

Mesenchymal cells lose this polarity, as well as the attachments to other cells and to the basement membrane, which makes it easier for these cells to move around

Julia Paxson

hold the cells together in the sheet, and create a waterproof seal around the fluid-filled blastocoel cavity. During gastrulation, cells that were originally part of the sheet of epithelial cells in the blastula are internalized. For this to occur, two different epithelial cell transformations can take place:

- First, many of these epithelial cells undergo epithelial-to-mesenchymal transition (EMT) as shown in Figure 4.7. EMT is essential in many processes—not only during embryonic development, but also wound healing, stem cell migration, and the metastasis of many different types of epithelial cancers. EMT is a multi-step process in which cells must reprogram their gene expression profile and change the proteins they produce, lose their affinity for their neighbouring cells, lose their apical–basal polarity, reorganize their cytoskeleton, change shape, and adopt a migratory phenotype which enables them to move along elements of the extracellular matrix. EMT involves the loss of the protein complexes that make up the tight and adherens junctions holding neighbouring epithelial cells together, as well as construction of cell-wide networks of actin filaments that will form a type of flexible skeleton that the mobile mesenchymal cell can use to move across the extracellular matrix.

- Second, some of the epithelial cells change shape. Transforming the cell shape from a cube to a wedge alters the shape of the sheet, as the cells pull on their surrounding neighbours. This process is known as apical constriction. It occurs in the endoderm during sea urchin gastrulation, and is also a critical part of neural tube formation in vertebrates.

Where are we now?

At this point in embryo development, the single fertilized egg cell has been transformed into a three-dimensional structure with complex populations of cells. These cells are progressively becoming more specialized, based on the positional information that they receive from their neighbours. But, by the end of gastrulation, there is still not much that distinguishes a worm embryo from a human embryo. Both are elongated, bilaterally symmetrical embryos, with a head and tail, belly and back, three germ layers, a digestive tract, and a nervous system that really hasn't even started to develop yet. It is in the next stage of embryo development, during organogenesis, that the differences between different animal species become more apparent.

Organogenesis: moving cells around

Organogenesis is the stage of embryo development after gastrulation, where cells in the three germ layers coordinate to form the internal structures of the animal and adopt their final identities. During organogenesis in vertebrates, the neural tube forms, organs such as the heart, lungs, and digestive tract develop, and limbs start to grow away from the main body axis. Although our DNA encodes the blueprints to create the necessary differences from the moment of conception, it is only at this stage of development that we start to become functionally humans instead of worms.

One of the most important events that occurs in all vertebrates is development of the central nervous system—the brain and spinal cord. In this section we will explore the complexities of developing different aspects of the central nervous system along the length of the anterior–posterior (head–tail) axis. The central nervous system (the brain and spinal cord) develops from ectoderm, whereas the protective bony structures surrounding it develop from mesoderm. In both cases, development is coordinated under the influence of regional patterning that is used to coordinate the correct functional specializations across multiple germ layers. For example, the brain must be surrounded by the bony skull, whereas the spinal cord must be surrounded by the articulating spinal skeleton, including ribs in the correct locations. Important among these mechanisms is the use of Hox genes in regional patterning, which we touched on briefly in Chapter 1 and which we will explore in more detail in Chapter 5.

Figure B All bilaterian animals, as well as Cnidarians (jellyfish and corals) and Ctenophores (comb jellies), have neural networks. However, Poriferans (sponges) and Placozoans (*Trichoplax*) do not, even though they have a more recent common ancestor with bilaterians and Cnidarians than the Ctenophores. A long-standing question has been—did the sponges and their relatives lose their neural networks, or did comb jellies develop their neural networks independently?

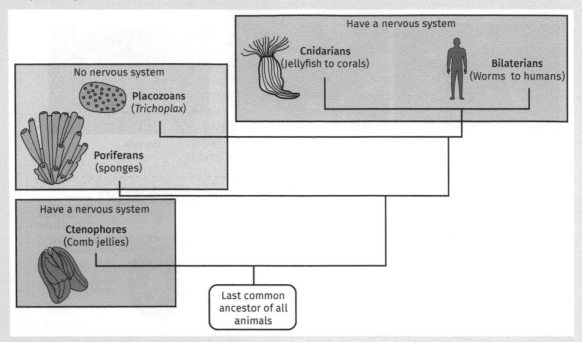

molecular patterning processes which specify the condensed nerve cords in different animal clades, as well as the use of different neurotransmitters in the neural systems of ctenophores compared to other animal clades.

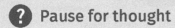 **Pause for thought**

If it is true that neural networks have evolved independently multiple times, this suggests that neural networks confer a huge evolutionary advantage. Conversely, as

with any specialized organ, neural networks are energetically expensive. A few animal species, such as the sponges, survive without a neural network and have either never evolved or lost these expensive structures. This story is a fascinating example how embryo development and evolution work cooperatively to generate phenotypes that confer advantage (whether in overall fitness or in sexual selection), and to lose phenotypes that are unnecessary. How might you apply this framework to think about evolutionary retention and loss of specific phenotypic features in other species?

Making a central nervous system

In any animal, the central nervous system is a collection of neurons and surrounding support cells that communicate together, receiving information from different parts of the body, integrating that information and providing a response. This is in contrast to the peripheral nervous system, which has a more limited role in providing information to, or receiving information from, the central nervous system.

The structure of the central nervous system varies depending on the animal but a hallmark of the neurons in the central nervous systems of many animals is their capacity to integrate information from multiple sources. Neurons themselves are remarkable cells. They are extremely specialized, with specific parts that enable them to receive (dendrites), integrate (axon hillock), and provide (axons) information. The information itself is communicated primarily through a combination of electrical signals (within a neuron) and chemical signals (between neurons). As with all other differentiated cells in the embryo, neurons are formed from the division and differentiation of neural stem cells. These stem cells originate from ectodermal stem cells at the end of gastrulation (see Figure 4.8).

How are the brain and spinal cord generated?

After gastrulation is complete, the elongated three-dimensional embryo begins the process of organogenesis, including neurulation. In vertebrates, both the brain and spinal cord begin as a continuous simple tubular structure that is created from another feat of embryo origami. The flat sheet of ectodermal epithelium that lies along the dorsal ridge of the embryo folds up on itself to

Figure 4.8 (a) and (b) Neurons and other neural support cells (glial cells) are derived from neural stem cells, which in turn come from ectodermal stem cells after gastrulation. (c) After the neural tube has formed, the neural stem cells start to divide. (d) After the stem cells divide (1), the daughter cells make their way to the outside edge of the neural tube (2), where they differentiate to form neurons and extend their axons towards their targets (3). (d) Neurons are very specialized cells.

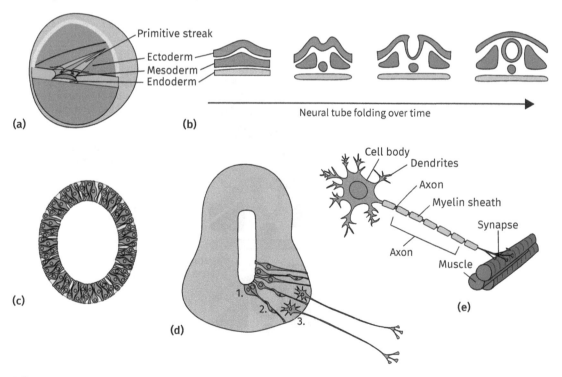

Julia Paxson

The development of the vertebrate brain

Along the anterior–posterior axis, the neural tube is rapidly divided into several different anatomical regions—most obviously forming the distinction between the brain and the spinal cord. We can visualize the adult brain as a series of interconnected fluid-filled cavities known as **ventricles**. Cerebrospinal fluid (CSF) is produced in the ventricles and provides protection and nutrients for the brain, which surrounds the ventricles. In the developing brain, a series of fluid-filled vesicles are formed and then gradually refined through a variety of bends, constrictions, and expansions. To form these fluid-filled bulges, the neural tube at the back of the brain transiently blocks, sealing off the brain and allowing rapid expansion of the brain vesicles as they fill with CSF. After initial expansion, the blocked region of the neural tube reopens to allow the flow of CSF throughout the spinal cord.

Neural stem cells in the brain initially form a single layer that spans the width of the neural tube. However, these neural stem cells quickly proliferate to form several different types of cells in a process known as **neurogenesis**. Newly created neurons populate different layers of the developing brain by migrating away from the inner surface of the neural tube (known as the **ventricular zone**) and out towards the outer surface (known as the **pial surface**). This is summarized in Figure 4.11. Some neural stem cells are also retained in the adult. In mammals, and mice in particular, recent research shows that these neural stem cells keep proliferating in some parts of the brain (for example the hippocampus and olfactory lobes). In mature animals, the neocortex—the part of the cortex in

Figure 4.11 The complex structures of the human brain are produced by the process of neurogenesis.

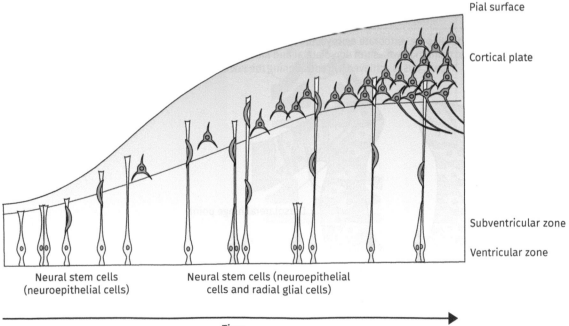

Pial surface

Cortical plate

Subventricular zone

Ventricular zone

Neural stem cells
(neuroepithelial cells)

Neural stem cells (neuroepithelial
cells and radial glial cells)

Time

Julia Paxson

mammals involved in sight and hearing—makes up the 'grey matter', coloured by the cell bodies of the neurons that reside there. The grey matter surrounds the deeper 'white matter' of the cerebrum, which consists of the axons that form connections between different parts of the brain.

Scientific approach 4.1
Neurogenesis post-embryogenesis

Until recently, scientists thought neurogenesis was a purely embryonic phenomenon. However, we now understand that, at least in certain areas in the brain and in the species most extensively studied, pools of adult neural stem cells are retained and used throughout the life of the animal. Adult neurogenesis seems to occur in specific regions of the brain, including specific regions of the lateral ventricles and the hippocampus (see the 2017 article from *Brain Research* shown in Figure A). As with other stem cells, adult neural stem cell populations shift between quiescence, proliferation, and differentiation. These shifts are sensitive to changes in signalling depending on cues from injury and environmental stimuli. For example, in the rodent hippocampus, the continual production of new neurons is associated with adult learning and memory. How do we know? Proliferating neurons can be visualized by the uptake of the thymidine analogue BrdU. When experimentally introduced into the brain, BrdU is incorporated into replicating DNA strands anywhere that thymidine would be used. BrdU in the replicated DNA can then be visualized using a fluorescently labelled antibody. When we look at the brains of rodents trained to complete a particular physical task, retention of BrdU-positive proliferating cells in several areas of the brain were increased compared to rodents that did not receive the same training (see Figure B).

It is worth noting that this data does not consider the origins of these proliferating cells, nor does it predict how similar events might occur in humans. To perform this

Figure A Papers are published regularly in the journal *Brain Research*.

ScienceDirect

Brain Research
Volume 1654, Part B, 1 January 2017, Pages 95-101

Review
Mental and physical skill training increases neurogenesis via cell survival in the adolescent hippocampus

Gina DiFeo, Tracey J. Shors

Show more ⌄

and non-coding RNAs that can alter post-transcriptional RNA processing and translation. During neurogenesis, for example, all three of these types of epigenetic modifications are employed at different points along the transition from neural stem cells to one of several different types of differentiated cells.

In this chapter, we have explored a few of the mechanisms that embryos employ during their transformation from a single totipotent stem cell to a three-dimensional organism. First, we discussed common molecular strategies that are used in inducing cells to adopt a particular cell fate. Then, we focused here on the ways in which cells move relative to one another during both gastrulation and neurulation to create new three-dimensional structures within the embryo. In the next chapter we will focus on additional processes that occur concurrently during organogenesis, including forming body segments and forming limbs.

Chapter summary

- Stem cells' progeny become committed to a particular cell fate through combinations of two basic mechanisms. First, cells can be induced to a particular cell fate because of maternally derived factors (often mRNA) segregated to a specific region of the mature egg or early embryo. Second, cells can be induced to a particular cell fate through cell signalling or environmental cues. In humans, the second type of mechanism appears to be more common.

- Cell specification is a sequential process. Often cells are first specified by germ layer (ectoderm, mesoderm, endoderm) and then also by position relative to one or more body axes (anterior–posterior or dorsal–ventral).

- Gastrulation is the process by which cells specified to one of the three germ layers move relative to one another. Mesodermal and endodermal cells move to the interior of the embryo and the ectodermal cells expand to cover the entire exterior of the embryo. During gastrulation, many cells change phenotype from epithelial to mesenchymal to enable them to move as individual cells. Cells can also move in sheets by changing the shapes of adjacent epithelial cell types to buckle or bend the entire sheet of cells.

- Neurulation is the formation of the central nervous system during organogenesis. The process of neurulation begins before or at gastrulation when the neuroectoderm is specified at the dorsal aspect of the embryo, often associated with low BMP signalling.

- After gastrulation is complete, the neural tube forms when the two lateral aspects of the neural plate fold up towards each other, eventually fusing together. The two new edges of the epidermis (lateral to the neural tube) also fuse to form the skin over the back of the embryo and overlying the neural tube.

- Neurogenesis is the process of neuron specification from neural stem cells. It occurs along the length of the neural tube, but neural stem cells are particularly prolific in the nascent brain.

- Differentiation of neurons from neural stem cells involves several different types of mechanisms, including asymmetric division patterns that lead to unequal segregation of PAR proteins and different levels of Notch signalling in the two progeny cells. In addition, differentiation of cells is often induced by complex patterns of epigenetic modification that can radically alter gene expression patterns within the differentiating cells.

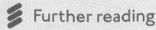

Further reading

The Dana Foundation (dana.org)

This is a great resource about the brain and neural anatomy that is very accessible and comprehensive.

Khan Academy: Overview of neuron structure and function (khanacademy.org/science/biology/human-biology/neuron-nervous-system/a/overview-of-neuron-structure-and-function)

This is another great resource covering neural anatomy and function.

Paridaen, J. T. M. L., and Huttner, W. B. (2014). 'Neurogenesis during development of the vertebrate central nervous system'. *EMBO rep* 15, 351. doi: https://doi.org/10.1002/embr.201438447

Discussion questions

1. Human embryos can be difficult to study, both ethically and because humans develop inside the uterus for most of development. What methods can researchers use to gain more insights into human embryo development? Consider researching different techniques, or model organisms that are easier to study and may provide some insights into mechanisms employed by human embryos.

2. Some diseases, such as that caused by the Zika virus, are known to affect the neurological development of embryos. Suggest ways in which these changes might be brought about at a cellular level in the embryo.

Figure 5.1 From flippers and flukes to wings and hands—limb development in vertebrate embryos offers a lot of variety based on the same underlying structural plan.

Anthony Short

All the right things in all the right places?

As you see in the earlier chapters of this primer, most animal embryos start off in very similar ways. But as you also know, the morphology and anatomy of the mature animal forms is very diverse. How does this happen—how do animal embryos end up with all the right things in all the right places to form creatures as different as a hummingbird and a blue whale? Despite how different these two animals look, it is also important to recognize that they are both also vertebrates, and therefore still have very similar underlying body plans, as well as very conserved embryonic strategies to build those body plans.

The two embryo development strategies we will focus on are body segmentation and the specification of regional patterns using Hox genes. Both strategies appear to be evolutionarily advantageous, and are broadly conserved among animal species.

- **Body segmentation** generates repetitive segments along the primary body axis. In some animals, specification of regional identity occurs alongside the generation of repeating segmental units along the primary body axis. This is most obvious in arthropods (think of centipedes!), but is also present in vertebrates (the human vertebrae, for example). By altering either the number or size of these segments, different body plans can emerge. For example, mice and snakes have vastly different numbers of these repetitive body segments, allowing for drastically different looking body plans (see Chapter 1).

- **Specification of regional patterns using Hox genes.** *Hox* genes are a broadly conserved mechanism used to drive regional patterning along the primary body axis across a wide range of animals (discussed very briefly in Chapter 1). As you will see further in this chapter, *Hox* genes act as master control switches to specify regional identities, and altering *Hox* gene expression patterns is another evolutionary strategy for producing diverse body plans.

Creating diverse and extreme body plans using body segmentation

Body segmentation occurs in several different phyla across the animal kingdom, including arthropods, annelids, and chordates (see Figure 5.2). There are fascinating questions about the functional, evolutionary, and developmental

Figure 5.2 The body segments in insects such as this dragonfly and in vertebrates such as this sea horse are easier to see than our own—but human embryos show clear segmentation too.

(a)   (b)

(a): Anthony Short; (b): GOLFX/Shutterstock

similarities and differences between body segmentation patterns and origins among these animals. However, here we will focus on the generation of repetitive body segments in vertebrates during embryo development.

Where and how does body segmentation occur in vertebrates?

Body segmentation in vertebrates occurs after gastrulation during a process known as **somitogenesis**. This is where blocks of tissue, known as **somites**, bud off from the rods of mesoderm (known as presomitic mesoderm) on either side of the developing neural tube (see Figure 5.3). Eventually these somites will be patterned to form several different components of the repetitive vertebral units along the embryonic spine, including the vertebral bones and muscles.

Figure 5.3 Body segmentation in vertebrates occurs through the process of somitogenesis, the sequential generation of mesodermal blocks of tissue known as somites on either side of the developing neural tube. Somites form from anterior to posterior from rods of presomitic mesoderm.

(a)   (b)

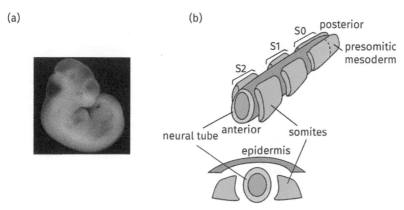

(a): Natalia Sinjushina & Evgeniy Meyke/Shutterstock; (b): Julia Paxson

the embryo is continuing to elongate during this time. Therefore, if the pulse of oscillating signal moves up the embryo more quickly, then more but smaller somites will be formed. If the oscillating signal moves up the embryo more slowly, then fewer but longer somites will be formed.

What are the consequences of somitogenesis?

As a result of the segmentation process, distinct blocks of defined mesoderm are created on either side of the neural tube (spinal cord). These blocks will eventually be shaped into the vertebrae and associated muscles that protect the spinal cord. The shapes of the vertebrae change along the anterior–posterior axis. For example, in mice, only the thoracic vertebrae at the level of the heart and lungs have ribs associated with them. In contrast, vertebrae in the neck do not. Human vertebrae vary from the atlas and axis pair that hold the skull and allow it to move, through the thoracic vertebrae that articulate with the ribs down to the heavyweight lumbar vertebrae where the muscles that hold the abdominal organs in place are attached.

How do the identically segmented mesodermal somites that are first formed get the positional information and molecular instructions that tell them what shape of vertebrae they should form? It starts with the same sorts of morphogen gradients that we have already looked at. Wnt and Fgf signalling are both highest in the posterior (tail) of the embryo and form morphogen gradients that decrease in concentration towards the head. In contrast, a signalling molecule called **retinoic acid** (RA) forms a morphogen gradient that is most concentrated at the back of the head and decreases towards the tail, as shown in Figure 5.4.

Figure 5.4 By the end of gastrulation in vertebrates, several different morphogen gradients have been set up across the anterior–posterior axis. Retinoic acid forms a morphogen gradient with highest concentration at the anterior end of the animal (which peaks at the hind brain), while both Fgf and Wnt form morphogen gradients with highest concentrations at the posterior end of the embryo. These morphogen gradients work to oppose each other and provide detailed positional information to populations of cells in all three germ layers. In this way, regional patterning can be coordinated throughout the entire embryo.

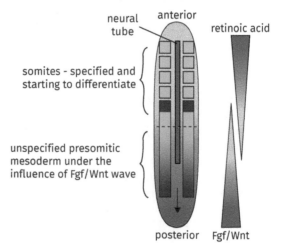

Julia Paxson

In combination, these morphogen gradients provide each somite along the anterior–posterior axis with positional information. This in turn causes expression of a specific class of molecules known as Hox genes that we will discuss next. The expression of unique Hox genes along the anterior–posterior axis provides all of the molecular information needed to generate vertebrae with the correct shapes in the right positions. Importantly, this system of positional information and Hox gene expression is used to pattern other tissues in all three germ layers along the anterior–posterior axis—it isn't just about the formation of the vertebrae.

Creating diverse and extreme body plans using mix-and-match regional patterning

How are regional patterning decisions made in the embryo?

Regardless of how, or even whether, an embryo develops a segmented body plan, positional information from the opposing morphogen gradients of Fgf, Wnt, and retinoic acid are also used in many animals to guide the formation of regional identities such as eyes, legs, or wings. Throughout the animal kingdom, regional patterning along the anterior–posterior axis is controlled by the sequential expression of a very special set of master control factors known as Hox genes that we mentioned in Chapter 1. Hox genes code for special transcription factors that drive regional pattern specification. These factors are themselves activated by positional information provided by the anterior–posterior morphogen gradients. Hox genes are highly conserved across every animal species that has been examined, both in gene sequence and in their regulation and expression during embryonic development. Hox genes are expressed sequentially along the primary body axis (head–tail), so that each region of the embryo is under the control of a different Hox gene.

If you looked at the case study in Chapter 1, you'll remember some of the key elements about Hox gene expression that makes them so very unique. Across the animal kingdom, Hox genes are expressed in sequence along the primary body axis, and exhibit a property known as collinearity. This means that the Hox genes are lined up on the chromosome in the same order that they are expressed along the primary body axis. This is illustrated in Figure 5.5—it is a remarkable attribute that is unique to Hox genes. In addition, Hox genes often function using the principles of posterior prevalence. This means that, although multiple Hox genes can be expressed within a given region in an overlapping manner, whichever Hox gene is usually expressed most posteriorly in the embryo will be the one that is active in patterning that particular segment. Therefore, the regional patterning in a particular segment of the embryo can be altered by either inducing or removing expression of the most posterior Hox gene expressed in that region.

How are Hox genes used for regional patterning decisions?

Much of the initial work on Hox genes was completed by T.H. Morgan and his students in the 1920s in *Drosophila*, as two clusters of genes that they named the Antennapedia and Ultrabithorax complexes, which we discussed in a case

Creating diverse and extreme body plans by altering limb development

Have you ever wondered what makes horses so fast and powerful when they run? Ever thought about the biomechanics of those beautiful long limbs and how very different they must be from our own limbs? But in fact, there isn't a lot of difference between us—just a small evolutionary improvisation on the same basic underlying tetrapod limb pattern that left us with five digits at the end of each limb and horses with only one. The term **tetrapod** refers to all those animals that have 4 legs, as well as those animals whose ancestors had legs but have since lost them (such as dolphins, whales, and snakes). Tetrapods evolved from ancestral lobe-finned fish, and the limbs of all tetrapods have the same developmental origins and the same homologous structures.

Think about the many different types of tetrapod limbs. Evolutionary changes in the ways in which tetrapod limbs are constructed have opened up a wide variety of different locomotion options. When horses are cantering, there is a point at which only one leg is weight-bearing. That means that for a racing thoroughbred horse, roughly 550 kg is pushing down on the bones and tendons of that one leg. What is even more remarkable is that the lower limb in the horse is the equivalent of a single digit. The superior running abilities of a horse is due to evolutionary adaptations that elongated and strengthened that single digit, while removing the other, superfluous digits. In fact, when you look at the anatomy of the limb bones in different tetrapods, you appreciate that they aren't radically different from one another. Instead, the differences are all due to small variations in the way that the basic tetrapod limb pattern has been interpreted (see Figure 5.7). But what huge differences those variations

Figure 5.7 Most tetrapod animals have all of these different limb components, but with specialized alterations depending on how they use their limbs.

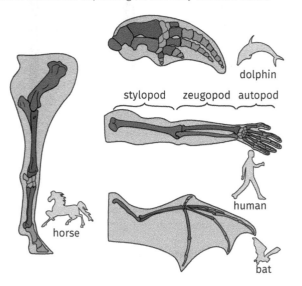

Julia Paxson

make—horses use their particular variations to becomes spectacularly fast runners, while bats use theirs to fly.

Understanding how the processes of growth and patterning occur within the microcosm of the developing tetrapod limb is fascinating and complex. Moreover, you can broadly apply the same molecular logic to an array of other biological processes. For example, tetrapod limb development is well-suited for investigating the regulatory relationships patterning mechanisms and signalling feedback loops because manipulating limb development is not lethal. Therefore, the effects of drastic loss-of-function and gain-of-function changes can be studied.

How is tetrapod limb development orchestrated?

Limb development involves the remarkable transformation of a small outgrowth of undifferentiated mesoderm covered with a thin layer of ectoderm into an anatomically complex, highly functional three-dimensional appendage. The limb mesoderm gives rise to over 60 bones (depending on the species), as well as associated muscles, tendons, and connective tissues. The limb also has a complex network of nerves and blood vessels, and it is seamlessly integrated with the main body structures of the embryo.

The tetrapod limb is a 3-D structure divided into three regions: proximal stylopod (humerus or femur), mid-axis zeugopod (ulnar/radius, or tibia/fibula), and distal autopod (wrist/ankle and digits) (see Figure 5.7). The digits are always arranged along the anterior–posterior axis (in humans for example the thumb is most anterior), and all tetrapod limbs have a dorsal aspect (back of the hand) and ventral aspect (palm). The milestones of tetrapod limb development include:

- **Initiation**. New limb buds are initiated through the intersection of several signalling pathways at specific points along the main body axis, causing a directed outgrowth of the mesoderm and overlying ectoderm. This process is directed by Hox gene patterning along the anterior–posterior axis.
- **Elongation**. The limb bud elongates by proliferation, condensation, and differentiation of mesenchymal cells to form the skeletal and muscular elements of the limb. This takes place in a proximal to distal direction, out from the centre of the embryo. At the proximal tip of the elongating limb bud there is a defined ridge of overlying ectodermal cells aligned along the dorsal–ventral interface, known as the apical ectodermal ridge (AER). Signalling from the AER drives proliferation of the underlying mesenchymal cells and elongation of the limb.
- **Arrest**. How long is the arm or leg? Final limb length is determined by a complex feedback loop between the proliferating mesenchymal cells, the overlying AER and an additional signalling centre known as the zone of polarizing activity (ZPA) located at the proximal-posterior aspect of the developing limb. When this feedback loop is fully functional, the limb will elongate, but when it is interrupted, limb growth stops.
- **Patterning**. Finally, the growing limb is provided with positional information in the proximal–distal, anterior–posterior, and dorsal–ventral axes, generating the regional identities that specify the stylopod, zeugopod, and autopod elements of the limb. At the distal end of the developing limb, a variable number of digits—for example 5 fingers or 3 toes—are formed. The digits emerge when programmed cell death or apoptosis in the distal digital elements removes the interdigital mesoderm.

Figure B The heading of the 2016 paper, where Leal and Cohn published their work into limb development in pythons and anole lizards.

Current Biology
Report

Loss and Re-emergence of Legs in Snakes by Modular Evolution of *Sonic hedgehog* and *HOXD* Enhancers

Francisca Leal[1,2] and Martin J. Cohn[1,2,3,4,*]
[1]Howard Hughes Medical Institute
[2]Department of Biology
[3]Department of Molecular Genetics and Microbiology
UF Genetics Institute, University of Florida, P.O. Box 103610, University of Florida, Gainesville, FL 32610, USA
[4]Lead Contact
*Correspondence: mjcohn@ufl.edu
http://dx.doi.org/10.1016/j.cub.2016.09.020

regions of genomic DNA that regulate transcription of Shh in the python ZPA. There is a regulatory region known as the ZRS to which transcription factors bind, thereby activating transcription of Shh. The sequence of this python Shh-regulatory region shows divergence from other tetrapods, and this could result in functional changes.

In contrast to lizards, python ZRS does not bind transcription factors strongly and therefore does not result in robust transcription of the Shh gene. How did the authors uncover this? They used a technique known as the **luciferase assay** in which a form of artificial DNA is introduced into cultured cells containing the ZRS sequence.

Instead of regulating expression of the Shh gene, the Shh DNA sequence is replaced by an artificially engineered luciferase reporter gene. How does this work? When the appropriate transcription factors successfully bind to the ZRS, it activates expression of the luciferase DNA sequence. When this luciferase mRNA is translated it generates luciferase (a fluorescent protein), and we can measure fluorescence of the protein product in the cultured cells. Compared to the lizard ZRS, binding of transcription factors to the python ZRS produced very little fluorescence, as shown in Figure C.

This work suggests that arrest of hind limb bud outgrowth occurs due to mutations that abolish specific transcription

Figure C This diagram illustrates how the artificial DNA constructs were generated. SV40 is a universal promoter, so if the ZRS is functional, the promotor should activate transcription of the downstream luciferase gene sequence. The lower graph compares luciferase activity (how much the cells were fluorescing when the anole lizard ZRS was used in the construct compared to the python ZRS).

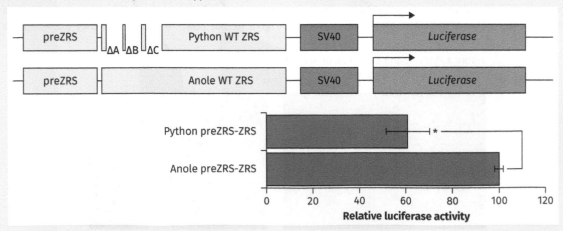

Leal, F., and Cohn, M. J. (2016). 'Loss and re-emergence of legs in snakes by modular evolution of *Sonic hedgehog* and *Hoxd* enhancers.' *Current Biology* 26(21), 2966–2973. https://doi.org/10.1016/j.cub.2016.09.020

Figure D In the model constructed by Leal and Cohn, the reduced functionality of the python ZPA is due to mutations in the Shh regulatory region known as the ZRS. This results in less transcription of Shh in the python and consequently more rapid decline of the ZPA–AER feedback loop—and no legs!

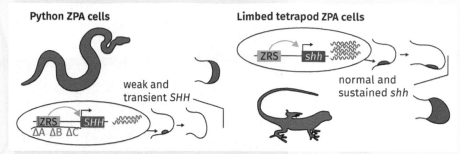

Leal, F., and Cohn, M. J. (2016). 'Loss and re-emergence of legs in snakes by modular evolution of *Sonic hedgehog* and *Hoxd* enhancers.' *Current Biology* 26(21), 2966–2973. https://doi.org/10.1016/j.cub.2016.09.020

factor binding sites in the ZRS. This is sufficient to cause early termination of outgrowth of the limb by termination of the ZPA–AER feedback loop and lack of Fgf8 expression, resulting in limbless snakes, as illustrated in Figure D.

❓ Pause for thought

In dolphin embryos, hind limb bud outgrowth is initiated in the same way as in other tetrapods, with a morphologically distinct AER that produces Fgf8. However, this signalling is transient and quickly lost in dolphins compared to other tetrapods. Similar to pythons, it appears that the primary defect in not in the AER, but rather in the ZPA. However, unlike pythons, in dolphins the ZPA is never functional and never produces Shh. What experimental tools might you use to provide support for this claim?

Limb regeneration: who can do it and how?

Now we have explored some of the concepts important in limb development, it is time to think about how these principles might apply to limb regeneration. Among vertebrates, salamanders are particularly well known for their regenerative capacity.

If a salamander loses a limb through amputation, it reconstructs an entirely new limb. Perhaps even more amazingly, only the correct missing structures will be generated. Regardless of whether the amputation is close to the body of the salamander or further away towards the foot, only the relevant missing structures are recreated, so that the final regenerated limb is indistinguishable from the original. This is truly a remarkable feat. After amputation, the remaining cells must have the ability to sense their original positional information and regenerate accordingly. Moreover, those original cells must either die, or de-differentiate so that new structures can be built in their place. Finally, large-scale cell proliferation, migration, and coordinated differentiation must occur in order to create all of the necessary new bones, tendons, nerves, blood vessels, and skin in the regenerating limb.

blocks presomitic mesodermal cells from being competent to respond. This is known as the 'clock and wavefront' model of somitogenesis.

- The number of somites formed can be altered by several different mechanisms, including changing the timing of the oscillating Notch signalling patterns. This can be used to create very diverse body plans in vertebrates.
- Once formed, patterning of these somites is dictated by Hox genes. Hox genes are highly conserved across the animal kingdom and display collinearity, so that these genes are expressed along the anterior–posterior axis in the same sequence that they are found along the chromosome.
- Changing the geographic distribution of Hox gene expression can also be used to dramatically alter vertebrate (and non-vertebrate) body plans. For example, the number of vertebrae with ribs is drastically different in snakes versus rodents.
- Limb development is another opportunity for creating diverse body plans. Elongation of the limb occurs through a positive feedback loop between a cap of epithelial cells at the distal end of the limb bud (known as the apical ectodermal ridge) and the underlying mesoderm. This feedback loop is further maintained by sonic hedgehog signalling from a sub-population of cells known as the zone of polarizing activity (ZPA). Eventually, inhibitory BMP signalling overwhelms the positive feedback loop and limb elongation is arrested.
- Limb regeneration occurs very successfully in some animals. Although later stages of growth and patterning appear very similar to limb development, there are initial differences. Chief among these is formation of a blastema at the distal tip of the amputated limb, rapid extension of nerves into this blastema, and probable recruitment of multipotent adult stem cells to populate the new limb.

 ## Further reading

Chal, J., and Pourquié, O. (2009). 'Patterning and Differentiation in the Vertebrate Spine'. In *The Skeletal System*, Cold Spring Harbor Monograph Series (Cold Spring Harbor, New York), 41–116.
Available to read in full at ResearchGate here: www.researchgate.net/publication/284478858_Patterning_and_differentiation_of_the_vertebrate_spine

Kaufholz, F., and Turetzek, N. (2018). 'Evolution of Segmentation: Sox enters the picture'. *eLife* 7:e41136. doi: https://doi.org/10.7554/eLife.41136

Keyte, A. L., and Smith, K. K. (2014). 'Heterochrony and developmental timing mechanisms: changing ontogenies in evolution'. *Seminars in Cell & Developmental Biology* 34, 99–107. doi: https://doi.org/10.1016/j.semcdb.2014.06.015
Available to read in full in preprint at the U.S. National Library of Medicine website here: www.ncbi.nlm.nih.gov/pmc/articles/PMC4201350

Discussion questions

1. This chapter is a short sampler of different concepts in developmental processes involved in somitogenesis, patterning, and limb development. Can you find the holes? What didn't we cover that you'd like to learn more about? For example, digit formation at the end of the limb involves patterning. Can you imagine how this might work? Think about it, then do some research to see if you can find out more!

2. Can you think of ways that you could use the principles outlined in this chapter to create other extreme body plans? For example, how might you use them to create the body plan of a dragon? What if the dragon had four legs and also wings?

Figure 6.1 Ageing is a fact of life in humans, but why does it happen—and can we stop it?

Ann Fullick

Rebuilding in regeneration: how does it happen and why can't we?

Tissue regeneration in animals encompasses a spectrum of processes and capabilities. The most masterful animal regenerators can regenerate their entire bodies (including their brains and all their other organs) from a small handful of cells. Other animals are able to regenerate limbs and other organs or—in the case of sea stars—bodies from amputated limbs! Finally, for those species, like ourselves, who are less accomplished in this area, regeneration merely includes rebuilding parts of organs (such as the liver), as well as tissues within an organ (such as intestinal epithelial lining).

Having already briefly explored limb regeneration in salamanders in the previous chapter, we will focus here on other types of regeneration, starting with planarian worms (flatworms), who are masters of whole-body regeneration. When cut into multiple pieces, planarian worms can correctly orient themselves to regenerate a head and tail on the correct ends of the amputated pieces. After three weeks, the regenerated worms are initially smaller than but functionally indistinguishable from the original, as shown in Figure 6.2. If you are most interested in questions related to human development and health, this might seem like a stretch. However, these little worms have a lot to teach us about how we might aspire to regenerate better ourselves!

Planarian worms are unassuming, but yet remarkable, creatures. They are non-parasitic, bilaterian, triploblastic flatworms. Because of their close evolutionary position relative to us, they demonstrate remarkable similarities to us in their molecular, cellular, and developmental processes. However, unlike us, most species of planarian worms retain large numbers of pluripotent stem cells after embryo development is complete. These stem cells (known as neoblasts) are scattered throughout the body and are key for successful whole-body regeneration (Figure 6.3). How do we know? If a mature planarian worm is irradiated, these rapidly dividing stem cells are wiped out, leaving the animal unable to regenerate. However, the introduction of even a single neoblast stem cell from another animal into the irradiated animal is sufficient to allow successful regeneration, after the stem cell has proliferated enough to repopulate the irradiated body. This is a truly stunning feat of stem cell biology!

Figure 6.2 Planarian worms, such as *Schmidtea mediterranea* have a remarkable capacity for whole-body regeneration. In the compilation of images shown here, a single worm is cut into 7 different pieces. Over the course of 10 days, each piece regenerates to form a new worm. Importantly each worm has a head in the same orientation as the original worm, suggesting that each cut section uses existing positional information to orient the regeneration process.

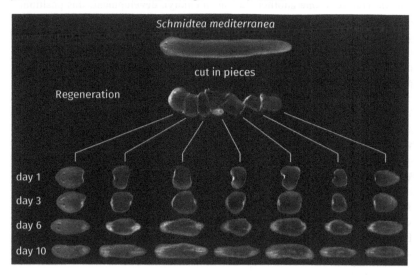

Jordi Solana

Figure 6.3 Neoblasts are found scattered uniformly throughout the planarian body, except for the pharynx (oral organ) as well as perhaps the brain (the head is on the left-hand side of the image, which is not actually at the same position as the pharynx in these animals!). In this top-down image, the entire population of neoblasts show up purple, while the actively dividing neoblasts are labelled in green. Even in non-regenerating worms, neoblasts are still needed for tissue homeostasis and therefore continue to divide actively.

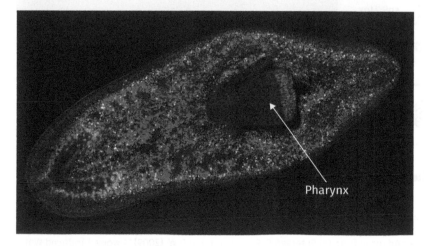

Image courtesy of Alex Lin and Bret Pearson

Scientific approach 6.1

Which signalling pathways are important for glial–mediated spinal cord regeneration in zebrafish?

The Journal of Neuroscience, May 30, 2012 · 32(22):7477–7492 · **7477**

Development/Plasticity/Repair

Fgf-Dependent Glial Cell Bridges Facilitate Spinal Cord Regeneration in Zebrafish

Yona Goldshmit, Tamar E. Sztal, Patricia R. Jusuf, Thomas E. Hall, Mai Nguyen-Chi, and Peter D. Currie
Australian Regenerative Medicine Institute, Monash University, Clayton, Victoria 3800, Australia

In this study, Yona Goldshmit *et al.* explore the role of Fgf signalling in glial-mediated spinal cord regeneration in zebrafish. First, they used **transgenic** fish with a **reporter construct** containing:

- a neural axon-specific protein promotor
- a green fluorescent protein (GFP) gene sequence

inserted into the fish genome. This means that any time the axon-specific gene is expressed, the GFP gene is also activated and all the neuronal axons light up with green fluorescence! Using this construct, the authors follow the progression of axon regeneration across a spinal cord bisection. When they combine this with immunofluorescence visualization of a protein that is specific for glial cells (GFAP), they see that glial cells form a bridge across the injury that allows regenerating axons to climb across and form connections with each other, shown beautifully in Figure A. Remember that immunofluorescence allows a researcher to use fluorescence-tagged antibodies to geographically locate specific proteins within prepared tissue sections.

Second, the authors provide correlative evidence that Fgf signalling is involved in glial bridging. They do this using *in situ* hybridization to show that the activity of several genes in the Fgf signalling pathway is strongly increased in bridging glial cells. Next, Goldshmit *et al.* used two different techniques to provide loss-of-function evidence that Fgf signalling is essential for the process of glial cell bridging. First, they used a chemical that inhibits Fgf signalling. Then, they generated another type of transgenic fish that contains artificial DNA with a dominant-negative version of an Fgf signalling component. This

gene was under the control of a heat-shock promoter—in other words, raising the temperature turns it on. When these fish are exposed to slightly higher than normal temperatures, the heat-shock promotor is activated. This in turn activates expression of the dominant-negative Fgfr1 gene. The result is transient inhibition of Fgf signalling. These results are also shown in Figure A.

Finally, the authors provide a little gain-of-function evidence by examining the effects of adding an Fgf signalling molecule to cell cultures of glial-like astrocytes from a mammal (marmoset) that is normally not able to regenerate neural axons using glial bridging. This experiment showed that in cell culture with Fgf signalling, astrocytes became much more bipolar (as shown in Figure B) and also better at migrating. Both these properties would hypothetically help form glial bridges after injury.

In this elegantly designed paper, Yona Goldshmit and her co-authors produce correlative, loss-of-function, and gain-of-function evidence that all supports the role of Fgf signalling in glial bridge formation during axon regeneration after spinal cord injury in zebrafish. Their model of this process is illustrated in Figure C. Unfortunately, it is proving hard to translate this work from animal models to people.

Pause for thought

How might this work translate into human medicine? Can you design experiments that you could perform to test whether Fgf signalling might help spinal cord regeneration in other animals? What limitations might you encounter?

Figure A The upper panels in this figure shows neurons fluorescently labelled green that are using bridging glial cells (fluorescently labelled in red) to affect regeneration. The bottom three images show the comparison of control (WT or wild type) regeneration compared to animals treated with either an Fgf signal inhibitor (SU5402) or transgenic fish in which Fgf signalling is inhibited in response to mild heat shock. Regardless of the method, if Fgf signalling is inhibited, so are the glial cell bridges.

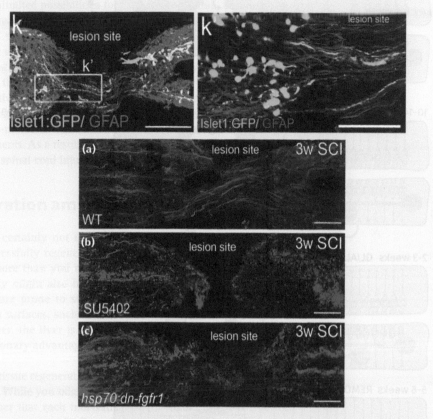

Goldshmit, Y., et al. (2012). 'FGF-Dependent Glial Cell Bridges Facilitate Spinal Cord Regeneration in Zebrafish'. *The Journal of Neuroscience* 32/22, 7477–7492

Figure B On the left is a cultured mammalian astrocyte displaying typical branched morphology. On the right, after Fgf treatment astrocytes display more bipolar morphology typically seen in the bridging glial cells.

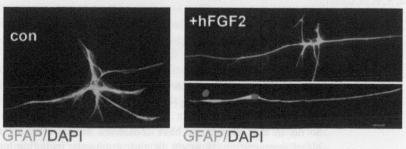

Goldshmit, Y., et al. (2012)

Figure 6.6 The hair regeneration cycle. In younger people, the anagen phase dominates and plenty of thick hair grows. With increasing age, the relative time spent in each phase favours telogen, eventually leading to hair thinning and loss.

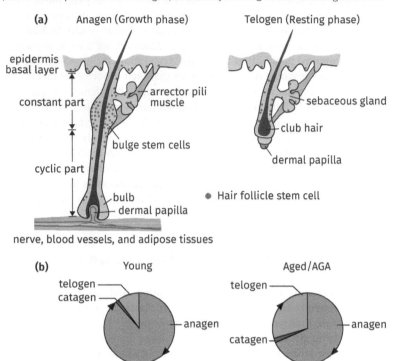

Modified from Ji, J., Ho, B.S.-Y., Qian, G., Xie, X.-M., Bigliardi, P. L., and Bigliardi-Qi, M. (2017). 'Aging in hair follicle stem cells and niche microenvironment'. *The Journal of Dermatology* 44, 1097–1104. https://doi.org/10.1111/1346-8138.13897

The loss of stem cell function in hair follicles with age is a common theme among adult multipotent stem cells in mammals. It leads to a general trend of reduced regenerative capacity with age. Therefore, when thinking about ways in which we can generate more functional regeneration in mammals such as humans, it is important not only to consider what type of permissive environment must be present (the types of stem cells and the signals they receive from the surrounding niche), but also how to prevent stem cell loss of function with increasing age.

What does all this science mean for humans in terms of understanding and improving our regenerative capacities? The more we understand about regeneration and tissue repair in our own bodies, and the more we understand about how other animals handle regeneration, the closer we will get to improving regeneration. It is not unrealistic to imagine that the combination of bioengineering, stem cell science, and regenerative medicine may soon combine to allow us to repair spinal cords, regrow limbs, and regenerate organs. This may seem far-fetched given everything we have discussed so far, but research into bio-artificial organs, stem-cell generated organoids, and bio-inspired scaffolds is fast-moving and shows promise in several different areas.

What is ageing—and who ages?

Perhaps these seem like obvious questions—isn't ageing obvious? Doesn't everyone age? The answer to both is a little more complicated than it seems. Ageing might appear easy to spot in humans (see Figure 6.1), but how would you define ageing in worms, for example? What about planarian worms with phenomenal whole-body regenerative capacity? Do they actually age? Is there a difference between the concepts of ageing and biological immortality? Can an animal not experience ageing, or age very slowly, but not be immortal?

There are, in fact, some constant markers of ageing that seem to be present across the spectrum of animals who age. The first is increased mortality. As the population of animals get older, they are more likely to die. Human mortality during adulthood increases roughly exponentially with age, following what's known as **Gompertzian mortality laws**. However, there are other organisms, including many mammals, that do much better than us. For example, the naked mole rat (*Heterocephalus glaber*) is the longest-lived rodent, with a maximum lifespan of over 30 years. Unlike us, naked mole rats are no more likely to die when they are old than when they are young. This is illustrated in Figure 6.7. Note, particularly, the comparison between naked mole rats and their close mouse cousins, which have much higher age-related mortality rates.

In addition to mortality, there are other factors that researchers consider in ageing. Most factors relate in some way to **cellular senescence**—the phenomenon by which cells enter an irreversible cell cycle arrest. Proliferating cells (such as stem cells) can undergo senescence in response to a variety of stressors and damage. Generally, senescence is an appropriate response in cells that might otherwise become neoplastic (cancerous), or in instances of repair where unchecked proliferation of cells such as myofibroblasts might lead to fibrosis. In young animals, the immune system is responsible for clearing the small numbers of senescent cells that are generated. However, as the animal ages, larger numbers of senescent cells are created (through mechanisms that we will discuss later in this chapter), and those cells are not cleared as well by the ageing immune system.

In addition to naked mole rats, giant tortoises (which can live roughly 200 years), Greenland sharks (which can live 200–300 years), and Quahog clams (including a clam that was documented post-mortem to be just under 600 years old) are all examples of animals with negligible evidence of ageing. In comparison to humans and other species that age, these species are characterized by the following:

- mortality rates that are consistently low across the lifespan
- low rates of cellular senescence
- low rates of diseases that are often associated with ageing and with increasing age-related mortality, such as cancer.

The longevity dilemma—reproduction or repair?

From an evolutionary perspective, tissue homeostasis and repair (known as somatic maintenance) are energetically expensive. In most animals, this must be balanced against reproduction, which is also energetically costly. Natural selection tends to optimize the balance between somatic maintenance and reproduction to maximize fitness for that population. In some animals, the

Case Study 6.1
Tumour suppressor gene P53—friend or foe?

P53 is a key tumour suppressor gene. It is activated in response to a variety of stress and environmental signals including DNA damage and oxidative stress, and results in cell cycle arrest. Importantly, cell cycle arrest is only important for cells that are still proliferating—those that are fully differentiated and no longer dividing will not be affected by this process. Depending on the type and extent of damage or stress signalling, this induced cell cycle arrest can be followed by successful DNA repair, apoptosis, or cell senescence if necessary, as shown in Figure A. Over 50 per cent of cancers in humans have a mutation in p53, which allows the cancerous cells to progress through the cell cycle despite the accumulation of DNA mutations or other cellular stressors that should result in cell death. For example, lab mice in which the p53 gene has been inactivated die extremely young due to tumour formation.

If we accept that all cells are equally likely to become cancerous, then larger animals (with more cells), or animals that live longer (and have more time to accumulate mutations), should be more likely to get cancer. Although this is generally true within species (for example larger dogs have higher rates of cancer than smaller dogs), it does not hold true between different species. For example, elephants have roughly 100 times more cells than we do, but significantly lower rates of cancer. This is known

Figure A The role of p53 in mediated cell proliferation and cell death.

Activation of p53 occurs in response to a variety of stress and damage signals and leads directly to arrest of the cell cycle. If repair is possible, the cell cycle may be resumed. If not, differentiation, apoptosis, or senescence may follow.

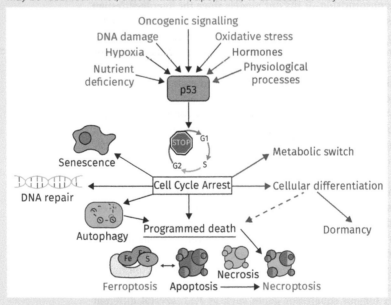

Moulder, D. E., *et al.* (2018). 'The Roles of p53 in Mitochondrial Dynamics and Cancer Metabolism: The Pendulum between Survival and Death in Breast Cancer?' *Cancers* 10(6), 189. https://doi.org/10.3390/cancers10060189 © 2019 by the authors. Licensee MDPI, Basel, Switzerland. This article is an open access article distributed under the terms and conditions of the Creative Commons Attribution (CC BY) license

as **Peto's paradox**. While lifestyle and environmental factors may play a significant role in some species, animals that are particularly large, or have particularly long lifespans have often evolved mechanisms to lower their cancer risk. For example, elephants have hedged their bets by having over 40 copies of the p53 gene (compared to our 2 copies). In the event that one or two copies of p53 become mutated in elephants, they still have many more non-mutated copies to rely on. In addition, elephants have evolved a cancer-killing 'zombie gene'—an extra copy of the Lif gene (which can act as an oncogene or tumour suppressor depending on several different factors). It is known as a zombie gene because it had first evolved to become non-functional, but was then later 're-animated' so it could cause apoptosis in cells when activated by p53 in response to DNA damage, as illustrated in Figure B.

Naked mole rats have taken a different approach, but with similar results. These rodents are not only exceptionally long-lived, but also very resistant to cancer formation. It turns out that their p53 protein has an extraordinarily long half-life, which means it might have superior tumour suppression activity. At the same time, high levels of p53 activity tend to inhibit stem cell proliferation and therefore reduce tissue repair in older animals. So, while p53 activity is necessary to lower the risk of cancer formation by tipping damaged cells towards senescence and death, high p53 activity can also lead to premature ageing, as illustrated in Figure C. Indeed, mice with enhanced levels of p53 activity show accelerated ageing that is associated with reduced stem cell proliferation and differentiation, as well as higher levels of cellular senescence. Therefore, we might also assume that the longevity of naked mole rats must involve other mechanisms to sustain cellular function, despite the high levels of p53 activity.

How does this translate in humans? One common point of variation in the p53 gene in people is at codon 72, where a single nucleotide polymorphism (SNP) results in an encoded amino acid at this location that is either an arginine (R72) or a proline (P72). The P72 version of the p53 protein is reported to be less effective as a tumour

Figure B Compared to humans, elephants have more cells, but less cancer. This seems to be due to repeated duplication of the p53 gene and a 'zombie' Lif gene that can be reactivated by p53 and induce cell death in response to damage.

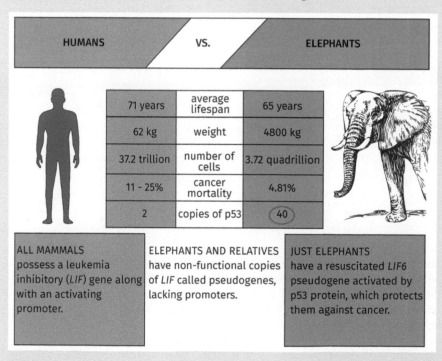

HUMANS	VS.	ELEPHANTS

71 years	average lifespan	65 years
62 kg	weight	4800 kg
37.2 trillion	number of cells	3.72 quadrillion
11 - 25%	cancer mortality	4.81%
2	copies of p53	40

ALL MAMMALS possess a leukemia inhibitory (*LIF*) gene along with an activating promoter.

ELEPHANTS AND RELATIVES have non-functional copies of *LIF* called pseudogenes, lacking promoters.

JUST ELEPHANTS have a resuscitated *LIF6* pseudogene activated by p53 protein, which protects them against cancer.

SITN / Krissy Lyon

Figure C The different roles of P53 in ageing and longevity. P53 is a necessary response to DNA damage that causes cell cycle arrest and reduces cancer formation. However, too much p53 activity can also lead to arrest of stem cell proliferation, reduced tissue repair and homeostasis, and reduced lifespan.

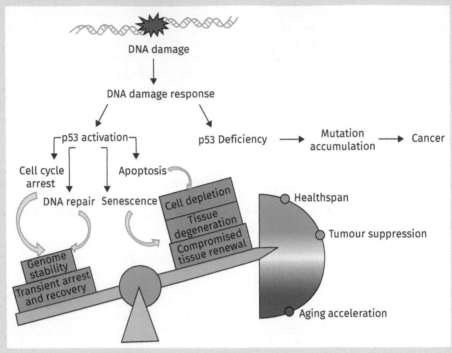

Ou, H.-L., and Schumacher, B. (2018). 'DNA damage responses and p53 in the aging process'. *Blood* 131(5), 488–495. doi: https://doi.org/10.1182/blood-2017-07-746396

suppressor, with less induction of cell death and senescence compared to the R72 version. However, despite the small increased risk of cancer in humans with P72, there is an overall increase in longevity, presumably because stem cell function is maintained for longer in ageing animals.

❓ Pause for thought

Can you imagine other ways to create an animal that avoids cancer successfully enough for it to be large and long-lived? What other factors might be important?

changes in differential gene expression without altering the DNA sequence in the genome. Common mechanisms of epigenetic regulation include changing DNA methylation patterns. This may affect histone modification, in which large areas of chromosomes are made inaccessible to the transcriptional machinery by being tightly wound around histone 'spools'. Alternatively, it may make individual genes more or less accessible by the transcriptional machinery in response to cysteine methylation patterns. There is now a lot of accumulated evidence to suggest that global genomic DNA methylation patterns are

Figure 6.9 Age-dependent epigenetic drift. Epigenetic regulation of gene expression patterns can be altered by regulating DNA methylation at a global level. DNA methylation patterns seem to be deregulated with age, leading to changing patterns of gene expression.

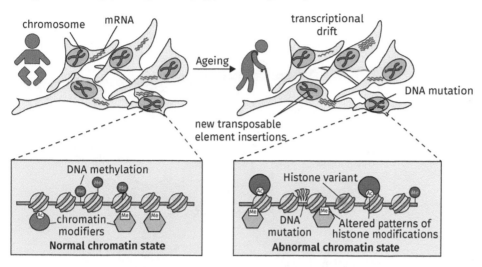

Pal, S., and Tyler, J. K. (2016). 'Epigenetics and aging'. *Sci. Adv.* 2, e1600584. doi:10.1126/sciadv.1 600584. © Sangita Pal & Jessica K. Tyler. Reproduced with permission

deregulated with age, as illustrated in Figure 6.9. The cells become less adaptable to internal and external signals, leading eventually to increased cell senescence and reduced cell plasticity. This phenomenon is known as **epigenetic drift** and is thought to overlay many of the other ageing processes that we will discuss throughout the rest of the chapter.

Why do we age, and can we (or should we) stop it?

Ageing is a complex process and not fully understood. We mentioned a few important considerations at the evolutionary level previously in the chapter, including optimizing fitness and antagonistic pleiotropy. We also mentioned the concept of epigenetic drift, which can affect many different aspects of cell physiology and function. Here, we will consider some important cellular mechanisms of ageing. However, there is a large and growing field of literature to dive into for more information and to see how all these themes intertwine.

First, let's consider some measures by which ageing is assessed at a cellular level. Previously, we briefly touched on the concept of cellular senescence, and this is certainly one of the foundation pillars of ageing. Cellular senescence can be induced by a number of intrinsic (inside the cell) and extrinsic (outside the cell) factors (see Figure 6.10). Among these, epigenetic drift can be measured holistically in reference to younger animals in the same species. Other causes include telomere erosion, irreparable DNA damage, oxidative stress secondary to mitochondrial dysfunction, and proteostasis dysfunction (problems with how proteins are made and appropriately destroyed).

Figure 6.10 Cellular senescence can be induced by changes in epigenetic regulation, telomere erosion, DNA damage, and mitochondrial dysfunction among others. In turn, cell senescence leads to stem cell exhaustion and is associated with ageing.

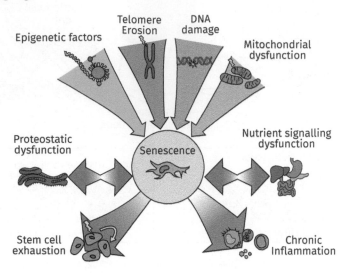

McHugh, D., and Gil, J. (2018). 'Senescence and aging: Causes, consequences, and therapeutic avenues'. *J Cell Biol* 217(1), 65–77. doi: https://doi.org/10.1083/jcb.201708092

Why does DNA repair capacity change with age?

The repair of both nuclear and mitochondrial DNA is a vital aspect of cellular function. Your DNA is compromised all the time by exogenous and endogenous processes, including ionizing radiation, UV radiation, and DNA replication mistakes, which must be repaired. However, the efficiency and extent of DNA repair is different across species, and also generally declines with age. As a rule, animals with better DNA repair capacity tend to live longer, as demonstrated in the graph in Figure 6.11. Larger animals with a higher total number of cells that might need to have their DNA repaired tend to have evolved more efficient DNA repair systems, and also live longer. Again, this might come back to optimizing fitness. Many small rodents are heavily predated and therefore experience heavy selection pressure to optimize fitness towards more reproduction and away from longevity. For those species, there is no advantage to evolving the most efficient DNA repair processes. The opposite is true for larger animals and those lucky small rodents that are not subject to the same heavy predation.

How might this information translate in humans? From the graph in Figure 6.11, you will be able to observe that we humans are fairly efficient at repairing our DNA (with direct correlation to our larger size). However, there are a collection of human diseases known as **progerias** where this might not be the case. People affected by progeria have characteristics of ageing that develop 8–10 times faster than other people. Among several different cellular processes that are affected by the mutations that lead to progeria is a tendency to increased DNA damage and/or less efficient DNA repair. Progerial diseases are devastating

Figure 6.11 After DNA damage, there are two broad categories of response. Either functional and proficient repair that ends in resumption of cell proliferation and tissue homeostasis, or lack of functional repair, leading to p53-induced cell cycle arrest, cell senescence, and ageing.

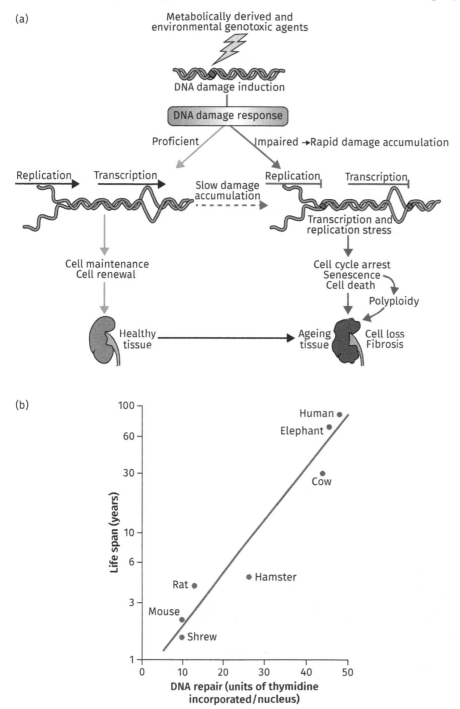

(a): Lans, H., and Hoeijmakers, J. (2012). 'Genome stability, progressive kidney failure and aging'. *Nat Genet* 44, 836–838. https://doi.org/10.1038/ng.2363. (b) The graph indicates the observed relationship between DNA repair capacity and lifespan

Figure 6.14 Although the total telomere length does not appear to correlate with lifespan, the rate of telomere shortening does. Animals that live longer have slower rates of telomere erosion.

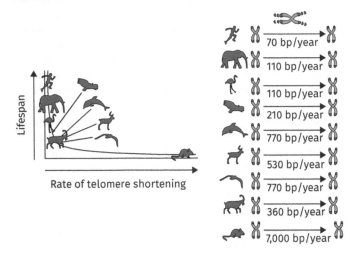

Whittemore, K., Vera, E., Martínez-Nevado, E., Sanpera, C., and Blasco, M. A. (2019). 'Telomere shortening rate predicts species life span'. *Proceedings of the National Academy of Sciences*, 116(30), 15122–15127. https://doi.org/10.1073/pnas.1902452116. Copyright © 2019 the Author(s). Published by PNAS

How does mitochondrial dysfunction affect ageing?

Mitochondria are well known for their role in energy harvesting and metabolism. However, they have additional roles in intracellular signalling, stress response, and stem cell function in both disease and ageing. Reactive oxygen species (ROS) and metabolic intermediates can act as intracellular signals that are particularly important for stem cell proliferation and differentiation. Mitochondria do not work as well as we get older. The root cause of age-related mitochondrial dysfunction is unclear, but it appears to include a combination of all of the mechanisms we have discussed so far—both nuclear and mitochondrial DNA damage and epigenetic deregulation. One gene that appears to be critical in this process is Sirt1. This codes for an enzyme responsible for mediating a host of important stress-sensing and response processes in the cell and in mitochondria. However, Sirt1 activity decreases with age, which appears to limit mitochondrial biogenesis and alter cellular stress responses.

How do these processes affect stem cell function with age?

Effective stem cell function is critical for tissue homeostasis and repair throughout the lifespan of an organism. However, epigenetic deregulation, accumulated DNA damage, telomere erosion, (and a host of other potential immune and inflammatory-related processes that we have not discussed) all limit stem cell function over time. Not only is the ability to keep dividing reduced, but the ability to adopt different cell fates is also limited with age. The result is less tissue repair, poor tissue homeostasis, and increased disease susceptibility. Collectively,

Figure 6.15 Age-related stem cell exhaustion: Proliferating stem cells experience age-related changes in both their ability to keep dividing and to differentiate. Factors contributing to stem cell exhaustion are shown in pink and orange.

Possible solutions to limit stem cell exhaustion are shown in purple. ROS refers to reactive oxygen species, which are a key element that contribute to stem cell exhaustion.

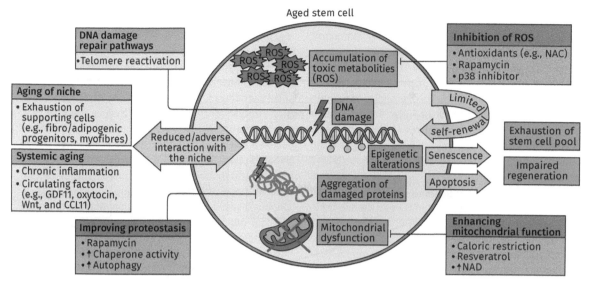

Oh, J., Lee, Y., and Wagers, A. (2014). 'Stem cell aging: mechanisms, regulators and therapeutic opportunities'. *Nat Med* 20, 870–880. https://doi.org/10.1038/nm.3651

this is known as stem cell exhaustion. You will notice in Figure 6.15 that stem cells must guard against many different avenues of stress and damage.

Should we stop ageing?

Can we stop ageing? Should we even try? There are a lot of commercial enterprises out there that try to convince us that we can and should! However, as you have seen, ageing is an extremely complicated phenomenon. While it is possible that ageing has not been actively selected for in our evolution (due to processes like antagonistic pleiotropy), ageing processes may be an evolved answer to avoid detrimental consequences such as cancer. There is currently a phenomenal amount of ageing research in model animal systems as well as some in humans. As we have noticed before, caution is appropriate when considering how to translate knowledge garnered from non-human animals into humans, since many animals have evolved with different considerations for optimized fitness and antagonistic pleiotropy. Having said that, there are some interesting ideas, including increased levels of Sirt1 as illustrated in Figure 6.16, which might be possible not only through drug therapy but through caloric restriction (CR), which we discuss more in The bigger picture 6.1.

In this chapter, we have explored two different themes that follow on from embryo development—regeneration and ageing. Both these processes rely, in part, on the functional attributes of stem cells and how these stem cells work to maintain the mature animal. By manipulating stem cell capacity, it may be possible for us to manipulate both the regenerative capacity and ageing processes in a particular species—even our own.

Figure B A variety of caloric restriction mimetics have shown promise at inhibiting ageing in many different organ systems and among many different animal species, including humans.

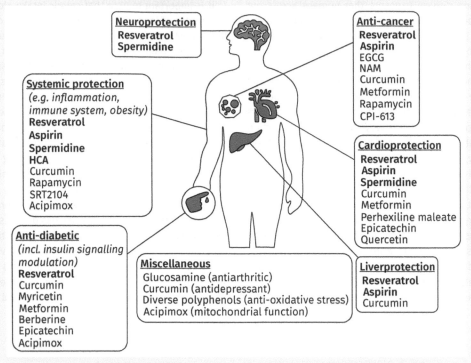

Neuroprotection
Resveratrol
Spermidine

Anti-cancer
Resveratrol
Aspirin
EGCG
NAM
Curcumin
Metformin
Rapamycin
CPI-613

Systemic protection
(e.g. inflammation, immune system, obesity)
Resveratrol
Aspirin
Spermidine
HCA
Curcumin
Rapamycin
SRT2104
Acipimox

Cardioprotection
Resveratrol
Aspirin
Spermidine
Curcumin
Metformin
Perhexiline maleate
Epicatechin
Quercetin

Anti-diabetic
(incl. insulin signalling modulation)
Resveratrol
Curcumin
Myricetin
Metformin
Berberine
Epicatechin
Acipimox

Miscellaneous
Glucosamine (antiarthritic)
Curcumin (antidepressant)
Diverse polyphenols (anti-oxidative stress)
Acipimox (mitochondrial function)

Liverprotection
Resveratrol
Aspirin
Curcumin

Madeo, F., Carmona-Gutierrez, D., Hofer, S. J., Kroemer, G. (2019). 'Caloric Restriction Mimetics against Age-Associated Disease: Targets, Mechanisms, and Therapeutic Potential'. *Cell Metabolism* 29(3), 592–610, ISSN 1550-4131

≋ Chapter summary

- Regeneration occurs at many different levels, from whole body regeneration in planarian worms to partial tissue regeneration in humans. Regenerative capacity depends on the innate abilities of stem cells, as well as signals from the niche surrounding those stem cells.

- In vertebrates, spinal cord regeneration after injury appears to depend on Fgf-induced formation of glial bridges, which can occur in some fish, but not in mammals.

- In humans, hair regeneration is a stem cell-mediated process that involves many of the signalling pathways (including Wnt and Fgf) that we've explored already. With age, the functional capacity of the hair follicle stem cells declines, and hair loss occurs.

- Ageing is a complex process that can be examined at many different levels that culminate in increased cellular senescence with age.
- Evolutionarily, selection processes may result in optimized fitness that prioritizes reproductive health over somatic maintenance.
- Antagonistic pleiotropy is another example of an evolutionary pressure in which gene variants may have opposite effects that result in higher fitness among young animals but decreased longevity.
- Epigenetic drift is the phenomenon of age-related deregulation of epigenetic regulation that can result in altered patterns of global gene expression with age.
- At a cellular level, unrepaired DNA damage, telomere erosion, and mitochondrial dysfunction can all lead to cellular senescence.
- Cellular senescence is particularly important in highly proliferative stem cells that are then unable to mediate effective tissue homeostasis.
- Although ageing processes may be reversible, we must be careful to recognize that ageing has evolved as one mechanism to avoid cancer. If we 'cure' ageing, we may also invite increased cancer formation.

 ## Further reading

Callier, V. (2017). 'A zombie gene protects elephants against cancer'. *Quanta Magazine*. (quantamagazine.org/a-zombie-gene-protects-elephants-from-cancer-20171107)

Cigliola, V., Becker, C. J., and Poss, K. D. (2020). 'Building bridges, not walls: spinal cord regeneration in zebrafish'. *Disease Models & Mechanisms* 13. doi: https://doi.org/10.1242/dmm.044131

Jones, O., Scheuerlein, A., Salguero-Gómez, R., *et al.* (2014). 'Diversity of ageing across the tree of life'. *Nature* 505, 169–173. doi: https://doi.org/10.1038/nature12789

Sánchez Alvarado, A. (2012). 'Q&A: What is regeneration, and why look to planarians for answers?' *BMC Biology* 10, 88. doi: https://doi.org/10.1186/1741-7007-10-88

 ## Discussion questions

1. Can you identify themes that are important in successful regeneration? If you try to stimulate regeneration in an animal that is not able to do so, where would you start?
2. What are the important themes in ageing? Explore ways in which animals with negligible ageing achieve this. In contrast, how might we increase our longevity by employing similar strategies?

GLOSSARY

Adherens junctions one type of cell–cell adhesion complex consisting of different types of proteins that is used to connect two neighbouring cells.

Adult stem cells See Stem cells

Ageing the phenomenon in which some (but not all) animals exhibit reduced organismal, cellular, and molecular functions as their chronological age increases.

Allogenic referring to cells taken from a donor of the same species that are used therapeutically in other individuals of the same species. (Contrast with *autologous*.)

Anagen the active phase of the hair follicle cycle, which can last up to 7 years in humans, while the quiescent (telogen) phase only lasts a few months in young humans.

Anencephaly a congenital variation in which the neural tube fails to close at the anterior (head) end of the embryo, causing an underdeveloped brain and incomplete skull.

Antagonistic pleiotropy an evolutionary theory that explains why some genetic variants persist despite some apparent disadvantageous effects in individuals.

Antennapedia a Hox gene in the fruit fly *Drosophila melanogaster*. A gain-of-function mutation of this mutation in *Drosophila* results in a homeotic transformation where the antennae are replaced by legs.

Anterior referring to the anterior–posterior body axis, where anterior is closer to the head.

Apical constriction a cellular phenomenon in which constriction of cytoplasmic filaments at the apical end of epithelial cells causes them to change shape from cuboidal to wedge-like.

Apical ectodermal ridge (AER) a specialized ridge of ectodermal cells that forms at the most distal (furthest from the centre of the body) tip of the developing limb bud.

Apoptosis a specific type of programmed cell death in which cells undergo characteristic changes such as blebbing, chromatin condensation, and DNA fragmentation.

Autocrine signalling a type of cell signalling pathway in which the cell that produces the signalling molecule (the inducer cell) is the same as the cell that responds to the signal (the responder cell). Other types of cell signalling pathways include juxtacrine, paracrine, and endocrine signalling.

Autologous referring to cells taken from an individual that are used therapeutically in that same individual. (Contrast with *allogenic*.)

Autopod the most distal region of the tetrapod limb (wrist/ankle and digits), along with the proximal stylopod and mid-axis zeugopod.

β-catenin effector protein for Wnt signalling, also used in cell adhesion complexes.

Blastula an early stage of embryo development in many animal species in which the embryo forms a hollow fluid-filled ball of cells.

Body axes the axes that define the overarching body plan. These generally include anterior–posterior (head–tail); dorsal–ventral (back–belly); left–right. Animals with limbs have an additional proximal–distal (near to the body–away from the body) axis.

Bone morphogenic protein (BMP) a paracrine cell–cell signalling pathway used in many aspects of embryo development, including dorsal–ventral embryo patterning, and limb development.

Cancer dysregulation of cells that enables them to divide uncontrollably and invade normal body tissue. If the cancer cells spread to distant parts of the body, it is known as metastatic cancer.

Cancer stem cells See Stem cells

Cell culture the experimental technique of growing cells outside the organism, often requiring specialized conditions that allow for controlled and replicable growth. Cells are often grown in culture to enable specific types of experimental manipulations, as well as for medical therapies.

Cell fate describes the future identity (or phenotype) of a cell or the progression of embryonic cells or stem cells as they adopt different sequential identities (or phenotypes) that will eventually lead to their final identity (or phenotype).

Cell proliferation the process of cell division that leads to an increased number of cells. In the embryo, this occurs as part of normal development. In the adult, this can occur through the controlled division of stem cells, or through the uncontrolled division of cancer cells.

Cellular senescence the phenomenon by which cells enter an irreversible cell cycle arrest. This is often associated with organismal and cellular ageing.

Central dogma of molecular biology an explanation of the flow of genetic information within the cell. Information in the genomic DNA is transcribed into mRNA (messenger RNA) in the cell nucleus. After the mRNA leaves the nucleus, information in the mRNA is translated to create a specific protein structure in the cytoplasm (or endoplasmic reticulum!).

Central nervous system the part of the animal nervous system that includes consolidated nerve cord(s) (the spinal cord in humans) as well as often an enlarged anterior section known as the brain.

Cerebrospinal fluid fluid produced by specialized cells in the brain ventricles that circulates throughout the brain and spinal cord.

Chemical signalling referring to the different types of communication used between cells.

Clades groups of organisms believed to have evolved from a common ancestor.

Cleavage the early cell division patterns that occur in the embryo in characteristics patterning specific to different animal species.

Collective migration referring to the ways in which groups of neural crest cells migrate together to common destinations.

Collinearity referring to the highly conserved phenomenon of Hox genes where they are spatially organized on the chromosome in the same order that they are expressed along the anterior–posterior (head–tail) axis in the animal embryo.

Correlative evidence See Scientific evidence (correlative, loss-of-function, gain-of-function evidence).

Dedifferentiation a process where cells that are fully differentiated and have a characteristic differentiated cell identity or phenotype become modified to be a less differentiated (more stem-like) cell type.

Differential gene expression the process in which only certain genes are actively transcribed within a given cell type, depending on many different factors including which transcription factors are present and active.

Differentiation (differentiate) the process in which cells express specific proteins that enable them to adopt a final cell identity or phenotype.

Ectoderm one of the three germ layers, the ectoderm forms tissues on the outside of the embryo such as the skin, as well as the neural system.

Effector protein in this book this refers to key mediators in signalling pathways. For example, β-catenin is a key effector protein in the canonical Wnt signalling pathway, mediating transcription of downstream target genes.

Embryonic organizer a group of cells in the early embryo that can function to induce different identities in surrounding cells. In many vertebrate embryos, the embryonic organizer induces dorsal identity in surrounding cells, regardless of their germ layer identity. Therefore, cells adjacent to the embryonic organizer that are part of the ectoderm would become specified as dorsal ectoderm, which eventually forms the neural tube.

Embryonic stem cells See Stem cells

Endocrine signalling a type of cell signalling pathway in which the cell that produces the signalling molecule (the inducer cell) is distant from the cell that responds to the signal (the responder cell) so that the chemical signal travels through the bloodstream (or other bulk convective system) between the two cells.

Endoderm one of the three germ layers, the endoderm forms tissues on the inside of the embryo such as the gut, liver, and lungs. The other two germ layers are the mesoderm and ectoderm.

Engineered stem cells See Stem cells

Environmental pressures (for evolutionary selection) this refers to any aspect of the external environment that can influence the survival of a population of organisms, thereby providing a pressure for evolutionary selection.

Enzymes specific proteins that are able to catalyse (accelerate) reactions in biological systems.

Epiblast the part of the mammalian embryo that will form the embryonic tissues, as opposed to forming the extra-embryonic tissues such as the chorion and amnion.

Epigenetic drift the well-supported theory that global genomic DNA methylation patterns are deregulated with age. The cells become less adaptable to internal and external signals, leading eventually to increased cell senescence and reduced cell plasticity.

Epigenetic regulation/modification this refers to broad mechanisms of affecting heritable changes in differential gene expression without altering the DNA sequence in the genome. Common mechanisms of epigenetic regulation include changing DNA methylation patterns.

Epithelial cell one of two broad cell phenotypes, the other being mesenchymal cells. Epithelial cells are characterized by being attached to one another in sheets, as well as to the underlying basement membrane.

Epithelial-to-mesenchymal transition (EMT) the molecular process by which cells change phenotype from epithelial to mesenchymal, losing their attachments to one another and becoming migratory.

Evo-devo a shortened term that refers to the field of evolutionary developmental biology, the comparative study of developmental processes in different organisms that enables inference of their ancestral relationships.

Evolutionary conservation of genetic function this refers to the observation that, despite very obvious differences in animal body plans, a large number of genes are conserved among different animal species, even those that appear distantly related.

Extra-embryonic membranes tissues in animals such as amniotes (birds, reptiles and mammals) that form from part of the early embryo to support embryo growth, including the chorion, amnion, allantois, and the yolk sac.

Fertilization the event in which two gametes (often sperm and egg) fuse to form a single cell embryo known as a zygote.

Fibroblast growth factor (Fgf) signalling a paracrine cell–cell signalling pathway that is used frequently during embryo development.

Fluorescent *in situ* hybridization an experimental technique in which nucleic acids (DNA or RNA) can be visualized in tissue sections using fluorescently tagged probes.

Gain-of-function evidence See Scientific evidence (correlative, loss-of-function, gain-of-function evidence).

Gastrulation the event during embryo development during which the germ layers (commonly ectoderm, mesoderm, and endoderm) are physically rearranged relative to one another such that they occupy their relative final positions, with the ectoderm on the outside, the endoderm on the inside, and the mesoderm sandwiched between them.

Gene networks in this book, these refer to hierarchical networks of genes that often work through transcriptional regulatory networks, in which the presence or absence of one transcription factor can influence the activation or inhibition of other transcription factors downstream in the network cascade.

Gene regulatory regions non-coding DNA sequences that can influence transcriptional activity

of the gene. For example, enhancer sequences (which can often be positioned at great distances from the actual coding sequence for the gene) can bind to transcription factors and form a hairpin loop such that the transcription factor complexes can then influence binding of the RNA polymerase complex to the promotor sequence for that gene.

Germ layers different functional groups of cells in the early embryo that form the basis for all differentiated tissue types. During gastrulation, these germ layers are rearranged relative to one another so that the ectoderm is on the outside, the endoderm is on the inside, and the mesoderm is sandwiched between them. Cells in the ectoderm germ layer form the skin and often (but not always) the nervous system. Cells in the mesoderm germ layer form connective tissues, bone, muscle, and blood cells. Cells in the endoderm germ layer form the linings of the gut and lungs, as well as the liver and pancreas.

Green fluorescent protein (GFP) a protein isolated from animals such as jellyfish that naturally fluoresces green, commonly used in experimental assays as a reporter for gene expression because the sequence for this protein can be artificially engineered to be expressed at specific times or in specific circumstances in the developing embryo.

Gremlin a bone morphogenic protein (BMP) signalling pathway inhibitor.

Gompertzian mortality laws theory stating that human mortality during adulthood increases roughly exponentially with age.

Hedgehog signalling a paracrine cell–cell signalling pathway that is frequently used in embryo development.

Heterochrony the evolutionary process of changing the timing of a developmental process (such as the timing of somite formation in somitogenesis), resulting in different types of animal body plans, e.g. creating more vertebrae in snakes compared to rodents.

Heterotopy the evolutionary process of changing the geographic distribution of a developmental process (for example Hox gene expression), resulting in different types of animal body plans (for example, changing the patterning of number of cervical and thoracic vertebrae in rodents, birds, and snakes).

Homeotic transformations transformations in regional patterning identities that can occur when Hox genes are mutated, e.g. the famous Antennapedia mutation in fruit flies wherein the antennae are replaced with legs.

Homologous refers to both homologous genes (genes in different species that have conserved

Phylogenomic studies studies that use information generated by comparing animal genomes across the phylogenetic tree.

Pial surface the outer surface of the developing brain.

Placenta a temporary structure that forms in mammals from the foetal chorion and the maternal endometrium to enable exchange of nutrients and waste products between the developing embryo and the maternal circulation.

Pluripotent See Potency (cell)

Positional information in the early embryo, positional information is provided by a variety of signalling pathways and molecular feedback loops to enable cells to become correctly specified and adopt the appropriate cell fates. Positional information is also a critical factor in successful tissue regeneration, such as limb regeneration or whole-body regeneration that occurs in some animals.

Post-embryonic growth and patterning processes that occur after embryonic development is complete, including metamorphosis, regeneration, and indeterminant growth processes.

Posterior referring the anterior–posterior body axis, where posterior is closer to the tail/anus.

Posterior prevalence a phenomenon present in Hox genes where the actions of a Hox gene located downstream on the chromosome (and therefore expressed in more posterior regions of the embryo) will inhibit the actions of a Hox gene located upstream on the chromosome.

Potency (cell) the capacity of a (stem) cell to replicate and differentiate into different types of cell fates. More potent stem cell types (totipotent and pluripotent stem cells) are considered more potent because they have a greater replication capacity and can often differentiate into a wider variety of cell types.

Primitive streak a transient embryonic structure that develops in most amniotes at the beginning of gastrulation. It denotes the geographic location at which cells on the surface of the embryonic epiblast start to migrate beneath this most superficial aspect and form the deeper mesodermal and endodermal layers of the gastrulating embryo.

Progerias a condition where characteristics of ageing develop 8–10 times faster than normal, usually significantly shortening lifespan. Cellular processes affected by the genetic variations that lead to progeria include a tendency to increased DNA damage and/or less efficient DNA repair.

Proliferative the capacity of cells to divide.

Regional identities different types of regional fates in the developing embryo, such as eyes, legs, wings, etc.

Regional patterning the process by which different regional fates are generated in the developing embryo.

Regeneration the process by which some organisms are able to completely recreate fully functional body structures without any scarring. At a tissue level, this may also refer to the capacity of some tissues to regenerate as part of normal tissue homeostasis, e.g., in humans, the epithelial lining of the intestines routinely regenerates as part of normal intestinal homeostasis.

Regulatory DNA specific DNA sequences to which different types of proteins, RNA, or other macromolecules can bind, resulting in either activation or inhibition of associated gene transcription.

Reporter construct an artificially engineered genetic construct. In this book, often referring to a genetic construct containing a gene-specific protein promotor associated with a reporter sequence that codes for a reporter such as a green fluorescent protein (GFP).

Responder cells in cell–cell signalling pathways, the inducer cell(s) produce/secrete the chemical signal that is biologically interpreted by the responder cells.

Retinoic acid in embryos, retinoic acid forms a paracrine signalling pathway that is used in many aspects of embryo development.

RNA polymerase the enzyme responsible for transforming information from genomic DNA into messenger RNA (mRNA) during transcription.

Scientific evidence (correlative, loss-of-function, gain-of-function evidence) different types of scientific evidence provide different degrees of support for a scientific hypothesis. Correlative evidence cannot prove causation and is therefore the weakest form of scientific evidence. Loss-of-function evidence can provide support for necessity, whereas gain-of-function evidence can provide support for sufficiency.

Segmentation in the developing embryo, different types of segmentation patterns can be generated.

Somatic cells all the cells in the embryo that are not germ cells (cells that have the potential to develop into gametes), or undifferentiated stem cells.

Somites repeating blocks of mesodermal tissues that develop on both sides of the neural tube along the anterior-posterior axis in vertebrates. Each somite can be further divided into the dermatome (forms the dermis), the myotome (forms epaxial muscles), and sclerotome (forms the vertebrae and associated ribs).

Somitogenesis the process of progressive somite formation that occurs in vertebrates from anterior to posterior along the developing embryo.

Sonic hedgehog signalling a paracrine cell–cell signalling pathway used in many aspects of embryo development, including from the notochord, ventral neural tube, and the limb bud zone of polarizing activity.

Spina bifida a condition that occurs in embryos with failure of posterior neural tube closure.

Stem cells cells that have the potential to form different types of adult cells. They include embryonic, adult, engineered (induced pluripotent stem cells (IPSCs)), and cancer stem cells and have different levels of potency—totipotent, pluripotent, or multipotent.

Stylopod the most proximal region of the tetrapod limb (humerus or femur), along with the mid-axis zeugopod and distal autopod (wrist/ankle and digits).

Telogen in humans, the quiescent phase of the hair follicle cycle, which only lasts a few months (at least in young humans). (Contrast with *anagen*)

Telomeres short repetitive sequences of nucleotides lined up at both ends of each chromosome. During genomic DNA replication, RNA primers initiate replication on the lagging DNA strand, so the new DNA sequence is a little shorter than the original sequence. Telomeres provide a non-coding buffer sequence that accommodates this by becoming shorter and shorter with each cycle of DNA replication.

Testable hypothesis a scientific concept that requires any hypothesis can be theoretically proved or disproved as a result of correctly designed and executed experiments.

Tetrapod animals with four legs, as well as those animals whose ancestors had four legs but have since lost them (such as dolphins, whales, and snakes).

Tight junctions one type of cell–cell adhesion complex that is used to connect neighbouring cells.

Tissue homeostasis the process of maintaining tissue integrity and steady state.

Tissue regeneration the process of recreating functional tissues after injury or during tissue homeostasis.

Totipotent See Potency (cell)

Transcription factors proteins that can control the rate of transcription of a specific gene by binding to associated regulatory DNA sequences and interacting with RNA polymerase.

Transcriptional regulation (activation and inhibition) the capacity to regulate transcription of specific genes through the interactions of transcription factors and DNA regulatory sequences.

Transdifferentiation the process by which some cells can change cellular identities or phenotypes.

Transforming growth factor B (TgfB) a paracrine cell signalling pathway that has multiple roles in embryo development.

Transgenic organisms that contain artificially engineered genetic constructs (such as reporter gene constructs).

Transmembrane receptors cell signalling receptors that span the cell membrane and are often used to transmit information from extracellular signalling molecules to and intracellular signalling cascade.

Triploblastic animals which have all three germ layers (ectoderm, mesoderm, and endoderm) in their body structure.

Tumour suppressor gene gene that regulates normal cell division and/or cell cycle progression and can thereby limit the potential for uncontrolled cell division seen in cancer.

Ultrabithorax Hox gene in the fruit fly *Drosophila melanogaster*. A loss-of-function mutation of this gene in *Drosophila* results in a homeotic transformation where the haltere structures in the third thoracic segment (T3) are replaced by wings usually found in the second thoracic segment (T2), resulting in a fly with two pairs of wings instead of one pair.

Ventricles the chambers of the developing heart, or the hollow, fluid-filled cavities in the centre of the developing brain.

Ventricular zone a transient embryonic layer of tissue containing neural stem cells, located at the ventricular (inner) surface of the developing brain.

Vertebrates animals that make up a large phylogenetic grouping based on the presence of a spinal column.

Wnt signalling pathway a paracrine signalling pathway with many roles in embryo development.

Zeugopod the mid-axis region of the tetrapod limb (ulnar/radius, or tibia/fibula).

Zone of polarizing activity (ZPA) a transient region in the developing tetrapod embryo limb bud, which expresses high levels of the paracrine cell signal sonic hedgehog.

INDEX

Note: page numbers with italic *b*, *f* and *t* indicate box, figure and table
see also Glossary